MANY LAYERS OF ECOCENTRISM

This book unveils the myriad streams of ecocentric thoughts that have been flowing through the human mind – in indigenous communities, in the wisdom of philosophers, in the creative expressions of poets and writers – sometimes latent, but sometimes more explicit. The strength of this book lies in the fact that it attempts to show that ecocentrism had not emerged suddenly as a distinct line of philosophical thought or found its place among the various normative approaches toward nature, but the seeds of ecocentrism had always been running through human societies. Thus, this book not only emphasizes the "unity of life" but also reveals the inherent unity of all hues of ecocentrism. The book adopts a multidisciplinary approach, which is essential to dwell on a topic like ecocentrism which permeates the domains of disciplines as disparate as science, philosophy, religion, normative ethics, myths and folklore, poetry, and literature, among others. Despite this eclectic approach, the book attempts to maintain continuity among the chapters and present these concepts in a simple form that will be easily accessible by readers from all conceivable backgrounds. This book would be useful to the students, researchers, and faculty from the fields of ecology and environmental science, philosophy, sociology, religious studies, and literature. It will also be an indispensable companion for all nature lovers, activists, and general readers interested in the emergence and evolution of environmental thoughts.

Abhik Gupta is a former professor of Ecology and Environmental Science and Pro Vice-Chancellor of Assam University, Silchar, India. He has an experience of over 35 years of teaching biology and environmental science in both undergraduate and post-graduate courses and has supervised over 25 doctoral theses. His research interests include biodiversity studies, environmental pollution, policy, and ethics. His recent publications include *Heavy Metal and Metalloid Contamination of Surface and Underground Water* (2020) and *Ethics and Biodiversity* (2021).

MANY LAYERS OF ECOCENTRISM

Revering Life, Revering the Earth

Abhik Gupta

Routledge
Taylor & Francis Group

LONDON AND NEW YORK

Designed cover image: Getty Images

First published 2024
by Routledge
4 Park Square, Milton Park, Abingdon, Oxon OX14 4RN

and by Routledge
605 Third Avenue, New York, NY 10158

Routledge is an imprint of the Taylor & Francis Group, an informa business

© 2024 Abhik Gupta

British Library Cataloguing-in-Publication Data
A catalogue record for this book is available from the British Library

ISBN: 978-1-032-60812-9 (hbk)
ISBN: 978-1-032-77076-5 (pbk)
ISBN: 978-1-003-48105-8 (ebk)

DOI: 10.4324/9781003481058

Typeset in Sabon
by Deanta Global Publishing Services, Chennai, India

To my mentor, Prof. R. George Michael (1935–2022), who introduced me to ecology.

CONTENTS

IN LIEU OF A PREFACE

The Rationale for This Book

This book intends to provide a brief overview of the different forms of bio-centric and ecocentric thoughts that can be found in all human communities. It also pitches the ecocentric and biocentric viewpoints against the ethical positions taken in anthropocentric (Anthropos: human; Kentron: center; meaning human-centered) approaches toward the treatment of nature. The terms biocentrism (Bios: life; Kentron: center; meaning life-centered) and ecocentrism (Oikos: house; Kentron: center; meaning nature-centered) are mostly used synonymously in this book, though the broader scope of eco-centrism by virtue of its inclusion of inanimate nature in its ethical concern is recognized. Admittedly, the author mostly takes sides with the greener views, though on occasions he relents to adopt an intermediate position between these two conflicting perspectives of human-nature relationships.

One of the major facets of human-nature relationships is whether such relationships recognize nature and humanity as two distinct entities, which can be termed as "a nature-culture divide". A second criterion is whether nature is regarded as an entity existing in its own right with its own intrin-sic value, independent of its use value for humans (Bogert et al. 2022). The question of values recognized in nature by humans has been discussed in Chapter 1. Within these two broad criteria, there could be shades of differ-ences in the perceived role of humans in nature. The Dutch philosopher Wim Zweers identified six possible roles in which humans could think themselves positioned vis-à-vis nature. Five of these positions are despot, the enlight-ened ruler, steward, partner, and participant. The sixth is the culminating point of "unity with nature" (Zweers 1994). Though considered as utopian, many individuals across cultures at different times in human history are believed to have achieved a state of unity with nature with no distinctions remaining between them and nature. However, in this book, I have followed the slightly modified version of Zweer's scheme advanced by another Dutch thinker, Petran Kockelkoren, whose classification comprises the four positions of master, steward, partner, and participant. Humans stand above nature in the role of master, and can exploit it to the extent desired by them. Steward stands above nature, but recognizes it as a gift of God or a precious heritage and is committed to take care of it, also because the humans of today are its guardians for handing it over to the next generation. Partner recognizes nature as its equal, and works in tandem with it, while participant envisages humans as part of nature in all possible sense – biological, cultural, and spiritual (Kockelkoren 1993, as cited in Van den Born 2007 and in de Groot 2010; personal interaction with Petran Kockelkoren; also see de Groot et al. 2011). As has been discussed in more detail in Chapter 2, the master and steward positions can be said to represent "strong" and "weak" anthropocentric positions, respectively. On the other hand, the ethical positions where humans visualize themselves as partners of and/or participants in nature could be identified with the biocentric or ecocentric worldviews. Many indigenous societies all over the world do not recognize any pronounced "nature-culture divide" and treat nature as having intrinsic value in their cultural practices. They also show a deep reverence for nature in their everyday activities and religious rituals. Besides, their cosmogonies, myths, and folklores reflect their love and respect for plants, animals, ecosystems, and other elements of nature. Both polytheistic and monotheistic religions also have biocentric/ecocentric streams of thoughts (see Chapters 5 and 6). Ecocentrism has been defined in several ways and in a range of perspectives. For instance, Lawrence Buell defined ecocentrism as

The view in environmental ethics that the interest of the ecosphere must override that of the interest of individual species. It is semi-synonymous with biocentrism in its antithesis to anthropocentrism, but whereas biocentrism refers specifically to the world of organisms, ecocentrism points to the interlinkage of "the organismal and the inanimate".

(Buell 2005, 137: quoted in Bellarsi 2009)

However, there are value-oriented definitions of ecocentrism as well, such as "non-human centered ontology that assigns moral value to non-human species and the environment" (Bogert et al. 2022). Some commonly available definitions of ecocentrism include "a philosophy or perspective that places intrinsic value on all living organisms and their natural environment, regardless of their perceived usefulness or importance to human beings" (URL 1); and "the philosophical assertion that all things have inherent value is known as ecocentrism" (URL 2). Thus, recognition of intrinsic value in both animate and inanimate nature is one of the hallmarks of ecocentrism. We can see that ecocentrism may appear in many hues and shapes: in the musings of a field biologist and ecologist during her or his wanderings through nature; in the surreal images of nature portrayed in poetry and art; in folk philosophical realizations in myths, folklores, and cultural practices; in religious and spiritual thoughts; in the penetrative wisdom of philosophers; or even in the understanding of the behavior of the molecules of life. This book uses the value-oriented understanding of the environment as the common thread to trace these varied ecocentric pathways.

Going back to the classification of human attitudes toward nature by Wim Zweers and others, the despot or master was the most commonly chosen during the spread of the colonies, and the global expansion and growth of industries, which caused serious disruptions in the environment accompanied by a rampant and widespread loss of biodiversity. However, with the realization of the damages done and the gradual spread of environmental awareness and activism, the one that is most followed now is that of the steward, which advocates conservation, recycling, and sustainability. One is skeptical, however, that this "have one's cake and eat it too" attitude may not work so well because many humans always tend to prioritize human needs and priorities over that of nature. It would also be difficult to find the obscure border between human need and greed. Not only that, despotic activities might hide behind the façade of a benevolent steward, and exploitation would continue unabated. A careful analysis of many developmental projects – despite the presence of EIA and other regulatory frameworks – is likely to reveal such underlying despotic tendencies. Many philosophers, scientists, environmental managers, and activists are of the opinion that neither scientific and technological advancements and innovations nor legislative and administrative measures solely by themselves would be able to strike at the root of the environmental crisis that the Earth is experiencing. Here, the question arises about the extent of prevalence and popularity of ecocentric inclinations among people – especially the youth. Environmental sociologists and psychologists have devised various scales and models to assess the extent of "pro-ecological" attitudes in the populace, among which the New Environmental Paradigm (NEP) scale (Dunlap and Van Liere 1978; Dunlap et al. 2000; Dunlap 2008; Hawcroft and Milfont 2010; Anderson 2012; Manoli et al. 2019) is perhaps the most widely applied all over the world. Earlier, Dennis C. Pirages and Paul R. Ehrlich advanced the concept of the dominant social paradigm (DSP), which represents the dominant societal worldview used to look at the world and its environment. This paradigm in the United States of the 1970s (and in the rest of the world) was characterized by a faith in growth and material abundance; the ability of science and technology to solve almost all problems; and the right of humans to subjugate nature (Pirages and Ehrlich 1974, 43: as cited in Dunlap 2008).

However, the continued onslaught of climate-induced disasters like floods, storms, and heatwaves, the rising sea levels and receding glaciers, the threatening presence of toxins in human body and food, plastic-choked waterways and even the seas, and the continued loss of biodiversity in the following decades is likely to have swayed public attitude toward a more bio- or ecocentric worldviews or at least toward weak anthropocentrism characterized by a steward-like attitude. Various surveys conducted with the NEP and other scales suggest that strong anthropocentric convictions appear to be on the wane among the students and youth in particular, and even in the adult population. Numerous surveys from different parts of the world involving school, college, and university students, teachers, senior citizens, and others indicate that more people now appear to hold weakly anthropocentric opinions, though it is not clear whether they subscribe to biocentric or ecocentric principles. Carina Lundmark also observed that the environmental ethic underlying the NEP scale "is 'shallow' rather than 'deep green'" (Lundmark 2007). Notwithstanding this critique, it may be noted that one question posed in the NEP scale concerns the equal rights of plants and animals to exist along with humans, which reflects a deep ecology principle. In most of the surveys, a large proportion of the respondents answered to this query in the affirmative, suggesting ecocentric inclinations. At the same time, many people also believe that high exploitation of resources is still feasible with the aid of increasingly sophisticated technology, especially bio- and nano-technologies, thus reinforcing belief in a weak anthropocentric position. The lack of efficacy of this compromise path is evident from the limited success of successive COPs to contain and resolve the climate change crisis. Even the promises of the Paris Agreement to keep the temperature increase within 1.5°C are now proving to be fraught with difficulties. On the positive side, among the Kunming-Montreal 2030 Global Targets of the 15th Conference of the Parties to the Convention on Biological Diversity held on December 7–19, 2022, Target 2 calls for ensuring that by 2030 at least 30% of the degraded areas of terrestrial, inland water, and coastal and marine ecosystems are brought under effective restoration and effectively conserved in order to protect biodiversity and ecosystem functions and services (Convention on Biological Diversity 2022). This perhaps marks the beginning of granting more rights to non-human organisms to exist and flourish in their own chosen ways.

All these developments suggest that if the exploitative DSP has to shift toward a more environmentally benign and biocentric or ecocentric paradigm, then there should be more discussions and debates among the general populace, especially the students and youth on the necessity of adopting a biocentric or ecocentric worldview. Humans are no strangers to such worldviews, since recognition of intrinsic value in nature was present in humans since the early days of their history.

A possible criticism of this book that it is an eclectic medley of information from diverse sources that appear to be far removed from one another can perhaps be defended on the ground that the ecocentric thought processes are also highly diverse and have flown along disparate pathways at different periods of human history at diverse geographical locations under distinct physical and socio-cultural settings. The author gratefully acknowledges the contributions of numerous natural and social scientists, philosophers, poets, writers, and artists toward the development of ecocentric ethics and philosophy; the indigenous communities for their ecocentric traditions; and the religious and spiritual leaders for their altruistic vision of a world where all forms of life would live and flourish together.

1

THE QUESTION OF VALUES

1.1 Introduction

The word value could be defined in a number of ways. Definitions given in the Merriam-Webster Dictionary include "the monetary worth of something", and in a broader perspective the "relative worth, utility, or importance". Thus, almost anything – a living organism or a non-living thing – could be assigned a value. Two types of values on which moral philosophers have debated and deliberated for long include the intrinsic or inherent value and the extrinsic or instrumental value. These values comprise the core of ethics, and subsequently have assumed paramount importance in environmental ethics as well.

1.2 Intrinsic and Instrumental Value

According to the American philosopher Christine Korsgaard, the commonest practice is to distinguish two types of "value of goodness", namely, "intrinsic" and "instrumental". Anything that has value because of "something else" is said to possess instrumental value. Common examples are tools, such as a screwdriver, a wrench, a pen, a pencil, money, etc. However, Korsgaard said that the exact opposite of this, that is "anything that is good for the sake of itself and not something else", is not the correct definition of intrinsic value. It would be more appropriate to say that anything that has intrinsic value has "goodness in itself", or in other words, it has "intrinsic goodness". The contrast of intrinsic value is not instrumental value, but extrinsic value or extrinsic goodness. The latter owes its goodness to an external source (Korsgaard 1983). This takes us to the question of finding

DOI: 10.4324/9781003481058-1

out about the things or the subjects and properties that can be thought to have intrinsic value. Zimmerman and Bradley (2019) in the Stanford Encyclopedia of Philosophy have referred to William Frankena (1973) to provide a long list of attributes which can be regarded as intrinsically good, and therefore, desired for their own sakes. These include conscious and active life, health, certain pleasures, happiness, contentment, truth, love, knowledge, wisdom, beauty, freedom, peace, security, honor, and several others. However, Frankena came up with a more detailed classification of values. He classified values into two main categories: moral and nonmoral values. Moral values are good on moral grounds, and there are no sub-categories. On the other hand, Frankena recognized six types of nonmoral values. Some things have "utility value", because their goodness depends on their usefulness for some purpose. For example, a pen or a pencil has value because of its utility, though if the nib is damaged or the lead is exhausted, it ceases to have any value. Digressing a little from the central issue of values, we can say that in the present context, this brings to light the relevance of the principle of reuse and recycle. If we replace the damaged nib or insert a fresh refill inside a ball pen, or have a facility where these could be melted and made into some other things, then their instrumental values are at least partly restored, and we temporarily avoid adding to the volume of nondegradable plastic waste. This principle applies to countless other items whose instrumental value we could prolong by reusing or recycling them. Frankena makes a distinction between utility value and "extrinsic value", the latter serving as means for obtaining or attaining good. We may think of money as belonging to this category, the possession of which may help us to obtain things that have utility value. Opposed to these values are "inherent" and "intrinsic" values, the former representing values that are inherently good, while the latter are good because they are intrinsically good, or are good by themselves. Frankena also recognized "contributory" and "final" values. The former arises when something contributes to something else to facilitate "intrinsically good life". Thus, we might say that clouds have contributory value, because they contribute water and shade, and, therefore, have contributory value toward a comfortable and prosperous life. Of course, we have to take into account the fact that a value may have different meanings and implications under different circumstances and to different subjects. For example, clouds formed in winter may be viewed in a negative light by many because they prevent the warmth of sunlight to reach the Earth, or cause rains that may be unwelcome in that season. On the other hand, the same winter rains will be greeted in arid regions by farmers, because it will save their crops from wilting. The last category of values – the "final values" – of Frankena are generally good. Though Frankena created so many sub-categories, the simplest approach is to club inherent and intrinsic values together as values that exist in their own right and do not need to be derived from any other value. The rest of the value categories are viewed as extrinsic or instrumental or utility/use value. Harold (2005) has defined extrinsic values as those needing justification by other values. In contrast to these, intrinsic values do not need to be justified by other values. All humans can be said to have intrinsic or inherent value in their own right, irrespective of their utility or their use as means to achieve some end. Hence, in other words, intrinsic values are not contributed by or derived from other types of values. On the other hand, extrinsic or instrumental values are derived from some other things or are justified by other values. A car has value because it can, say, transport us faster and in comfort. However, if one gets a still faster or a more spacious car, the value of the earlier car gets reduced.

We can also view the concept of values from the perspective of the German philosopher Immanuel Kant, who set forth what is often called the "Formula of Humanity" as one of the formulations of his "Categorical Imperative" – which he considered as the apex principle of morality. In this formulation, Kant asked to treat all humanity as an end, and not as a means to an end. For example, a car is a means to achieve some end, which in this case is reaching the destination faster or more comfortably, and a faster or more spacious car achieves that end more efficiently. On the contrary, the same cannot be said about the value of humans, who are ends in themselves. Of course, one can say that at many points of time in our history in many cultures, humans have been treated as objects or as means, and not as an end in itself. We can cite the example of slaves or bonded labor, who did not enjoy any right by themselves, and therefore, had a value that could be categorized as utility/instrumental/extrinsic. When slaves became old or infirm, they lost their value and were often unceremoniously dumped or neglected. Nevertheless, the image of humans that we normally have reflects the intrinsic value contained in them. If anyone flouts this directive to treat a fellow human as a means, then it goes against the principle of morality (Kerstein 2019).

Here I must say that the definitions and explanations of the different types of values given above are somewhat simplified, and do not cover all the subtle inconsistencies and variations that are bound to be there in the laying down of such complex and somewhat diffuse concepts. There is a lot of debate about the ambits of the different types of values, and perceptions differ among various moral philosophers and thinkers on their exact nature and delimitations. The discourse here is confined to the basic premises since it is not the purpose of this book to elaborate on the value debate but to understand it in the context of its relevance in understanding anthropocentric (human-centered) vis-à-vis biocentric/ecocentric (life- or ecosystem-centered) worldviews, which will be discussed in the subsequent chapters of this book.

1.3 Value(s) of Nature

Because values are concerned with the worth of an object or individual or any other entity, we can say that we treasure and protect those entities that have some sort of value. However, in the case of the objects having only instrumental/extrinsic value, these cease to be of any worth after their value is lost or expired. A life-saving drug, which has high extrinsic value, is disposed of as waste after its date of expiry. Again, a drug considered life-saving in an earlier period loses its value when superseded by a more effective modern compound. Thus, even extremely high extrinsic value assigned to an entity is also transient, and may be gone once the utility is over, or when it is replaced by a more useful entity. Domestic animals including pets are regularly abandoned when they become old or infirm or are no longer the subjects of love and affection of their owners. The indigo plant (*Indigofera tinctoria*) was once greatly valued as a source of indigo dye, and farmers in Bengal, India, were coerced by indigo planters to sow this crop, leading to a bitter and protracted political struggle termed as the "indigo revolt". However, the flourishing industry crashed with the success of some German scientists in producing synthetic indigo dye from coal tar toward the end of the 19th century (Asiaticus 1912; Indian Culture 2022). This caused a once valuable plant to be robbed of its value because only its extrinsic or instrumental quality was taken into consideration. It was subsequently termed as a "weed" – e.g., "agricultural weed", "sleeper weed", "garden thug", among others (Datiles and Acevedo-Rodriguez 2014) – in the Global Compendium of Weeds (Randall 2012). It is also named as an invasive plant in USA, and is, therefore, considered a useless and even harmful plant in human terms. However, the tables might turn again with a growing revival of interest in organic indigo for dyeing textile and hair. Indigo cultivation is gaining popularity and profitability in several states of India (Pachauri and Parker 2018). However, even in a highly commoditized world, humans are accorded intrinsic value. Great debates and discussions have taken place over the question of whether nature and its various components in the form of plants, animals, and microorganisms, along with the vast array of ecosystems that harbor these life forms, are to be accorded intrinsic value, or their value be merely limited to the extrinsic or instrumental type. While the philosophical debate around this issue had started much later, human relationships with nature are a part of the cultural portfolio of a given society at a particular period of time. In the early societies, this relationship was not so much reflected in philosophical treatises as it was in folklores and other oral traditions, religious practices and rituals, and in laws and regulations largely framed and prescribed by the respective communities.

1.4 Values of Biodiversity

Plants have played very significant roles in the development and evolution of human societies. The modern human species – *Homo sapiens* – evolved in the course of hominid evolution about 300,000–315,000 years ago in Africa (though there are a host of opinions on the timeline). Even in the hunter-gatherer communities, plant-based food items such as tubers, plantains, fruits, nuts, and others formed a substantial part of the diet, though meat formed a major constituent. The close association of plants with humans led to the practice of worship of trees and other sacred plants in many early societies. Similarly, animals and entire ecosystems such as forests, lakes, and rivers are still considered sacred and worshiped in many societies. Thus, recognizing intrinsic value in nature was prevalent in almost all early societies, though philosophical discourses on the values of nature could be said to have begun in medieval Europe only. These debates also take us to the question of anthropocentrism versus bio- or ecocentrism in relation to the question of intrinsic versus extrinsic values of nature and natural entities.

Peter S. White suggested that two types of values provide the basis for conservation of biological diversity. White clubbed extrinsic and instrumental values together to define extrinsic value as "a value external to the entity to be valued". In contrast to this, "intrinsic value is value inherent to the entity valued" (White 2013, 1280). He listed the drawbacks of using extrinsic valuing of biodiversity, which include measurement, scale, context, negative species value, and human subjectivity. Regarding measurement, it may be possible to set a monetary value on marketed or marketable biodiversity or biodiversity parts and products. However, there could be much more that is valuable in an indirect way and/or in terms of future potential. And the monetary value may also change with a change in scale or context. Scale is an important factor that can influence extrinsic valuing. White cites the example of the proposal to reintroduce Florida panthers in areas where it was eliminated. This move was supported by most people in Florida, though the rural communities vehemently opposed this action. Similarly, the context may alter the extrinsic value of biodiversity. For example, the recent COVID-19 pandemic greatly reduced the tourism value of hotels, homestays, transport operators, and other traders near the national parks and wildlife sanctuaries in India. Another problem is created by the negative values that some species or ecosystems have from a human point of view, leading to human actions such as predator control and draining of swamps. However, these same actions are detrimental to the other species and to the ecosystems that harbor them. In that sense, predators and even parasites and disease vectors have some positive role to play for other species and ecosystems. Predators keep prey populations under control, and disease vector life stages like mosquito larvae serve as fish food. Such conflicting negative values from a human viewpoint but positive values for other species create problems for extrinsic valuing of biodiversity. Human subjective opinions and preferences for certain species are the other sources of complication for extrinsic valuing. White contended that since intrinsic value is "absolute" and serves as an end and not a means to an end (Kant's categorical imperative), it does not have the problem of gradation of measurement as is encountered with extrinsic value. If these values are measured on a scale of 0 to 1, intrinsic value is always equal to 1, while extrinsic values could range from 0 to <1. However, the problem arises from the fact that the intrinsic value of biodiversity is almost always a form of hidden extrinsic value. Further, the intrinsic value of one species is "often in conflict with" that of another species. White raised questions about whether intrinsic value is manifested in the various elements of biodiversity such as "genes, individuals, species, ecosystems, landscapes" or its processes such as evolution, adaptation, population dynamics, integrity, health, resilience" (White 2013, 1283). Here, at one extreme is recognizing value in individuals. Hence, the question is which individuals or species are to be given intrinsic value? Many ethicists are in favor of granting intrinsic value only to sentient organisms (those capable of feeling pain), which is a more restrictive approach. On the other extreme, there are people who find intrinsic value in entire ecosystems. The issue is further complicated by the fact that biodiversity and the systems and processes such as ecosystems, energy and material flows, succession, evolution, etc. are so intricately linked that they cannot be easily separated from each other. White concluded that biodiversity possesses both extrinsic and intrinsic values, though intrinsic value "is the primary value system for life". Therefore, to conserve biodiversity on the basis of intrinsic value would conserve it in its entirety (White 2013, 1285). Contrary to the contentions of thinkers like White and many others, Karen Green argued that "most values in nature are objectively existing extrinsic values", though "the value of the biosphere" could be an exception (Green 1996, 32). Green referred to Korsgaard (1983) where she recognized instrumental value as a type of extrinsic value, the latter also including other types of extrinsic values such as contributive value and inherent value. Therefore, the value of nature (and biodiversity) could be found in such non-instrumental extrinsic values. Green cited certain examples to take this argument forward. For example, Cape Barren Geese (*Cereopsis novaehollandiae*) is a large goose found in the southern coast of Western Australia and in south-eastern Victoria. The numbers of this bird greatly dwindled in the 1950s, necessitating urgent conservation measures. This enabled the geese population to recover and increase in number, and consequently, the priority for Cape Barren Geese conservation was also reduced. Similarly, Green also cited the examples of the tall old-growth trees in Tasmania and the Queensland Daintree rainforest, and argued that if the Tasmanian forests were not severely logged, or the loss of the Queensland rainforest was not significant, then the value of an acre of Tasmanian forest or Queensland rainforest would not have been as great as it is today. We can also raise the issue of the nature of values in invasive plants or pest animals. It is to be understood that a plant becomes invasive or an animal becomes a pest only under certain conditions created by humans. Hence, an ethical responsibility of humans would be to prevent such situations to develop.

At this juncture, it may be relevant to point out that according to eco-centric and Deep Ecology principles, all forms of life – whether large or small, commercially important or not important, rare or abundant – possess intrinsic value. Therefore, the Cape Barren Geese, the Tasmanian tall trees, or the Queensland rainforest all possess intrinsic value, irrespective of whether these are endangered or dwindled or rare or are abundant and safe. According to Green, the extrinsic values of these entities go up or down depending on their status. On the other hand, their intrinsic values remain constant and absolute, as pointed out by White (2013). Accepting this position implies that intrinsic value has to be recognized in all forms of life, and not only in an organism or species that is categorized as "Critically Endangered", "Endangered", or "Vulnerable" in the "IUCN Red List of Threatened Species" (IUCN 2022).

In his "Duties to Endangered Species", Holmes Rolston, III pointed out that it would require an "unprecedented mix of biology and ethics" to achieve biodiversity conservation (Rolston 1985). Many powerful reasons have been advanced to justify the conservation of biodiversity, albeit with a strong anthropocentric core. Some early ethicists (e.g., Feinberg 1974; as cited in Rolston 1985) felt that saving other species is not a duty to that species *per se*, but is a part of human responsibility to the future generations of humans. Rolston also referred to the "rivet-popper" hypothesis of Paul and Anne Ehrlich (Ehrlich and Ehrlich 1981; as cited in Rolston 1985) where they compared the biological species to rivets in the "earthship in which humans are flying" (Rolston 1985, 718). A species going extinct means a rivet lost, and too many such losses are likely to send the earthship crashing. Rolston interpreted this hypothesis as suggesting that any species not serving as a rivet (provided there are any such species) would not have any value because humans are only interested in saving those species, the loss of which can cause a "crash". However, some of these "non-rivet species" could serve as resources, although only a fraction of the species has so far been screened for their properties or endowments useful to humans. Ehrlich and Walker (1998) clarified – albeit from a utilitarian point of view – the "rivet popper" hypothesis and the concept of redundancy in the context of their implications in biodiversity conservation and human dependence on biodiversity for the survival and well-being of humans. The rivet popper concept recognizes the existence of redundancy in ecosystems where redundant species may be compared to rivets in an aircraft. That does not mean that some of these species may be removed, because it is not known which species are redundant. According to the redundancy hypothesis, extinction of a few species may not be overtly harmful because species equivalent to those lost are likely to be there in an ecosystem. However, this knowledge does not give humans the license to remove species at will, because redundancy can serve an important role in the event of increased stress, and therefore, maintaining redundancy would increase "ecosystem resilience". The basis of all these arguments is that they only ascribe extrinsic or instrumental value to biodiversity. It is, therefore, ethically more appropriate to consider Rolston's contention that to protect the nonhuman species solely from a human-interest viewpoint, or in other words, only recognizing their extrinsic/instrumental value, maybe a prudent exercise, but it cannot originate from a sound moral standing. As Rolston said, "the challenge now is to learn interspecific altruism ... what is required is not prudence but principled responsibility to the biospheric earth". Rolston concluded that regarding "one species" (the human species) "as absolute" and valuing "everything else relative to its utility" is "something morally naïve" (Rolston 1985, 720, 726).

At an early phase of the anthropocentrism-ecocentrism debate, Donald Worster observed the dichotomy of the thoughts of two schools of environmentalists – one being the older anthropocentric school insisting on the relevance of an issue solely from the point of view of human interests. On the contrary, the (relatively) new group of ecocentric thinkers justified human moral obligations to nature as a whole. Among these latter were people like Schweitzer, Leopold, Thoreau, and others who believed that the moral responsibility of humans was not confined to the members of their own species, but was required to extend to the entire living community. Thus, it was not enough for humans to enjoy their rights and be liberated, but it was "nature's turn to be liberated" (Worster 1980, 45). Worster nurtured the hope "to salvage another relic from the age of domination: an instrumental view of nature by which everything in this complex, living world is to be subordinated to man's ends" (Worster 1980, 45). If this ethical position is endorsed, then nature cannot have any value in itself, because it is merely meant for fulfilling human goals and aspirations.

Michael E. Soulé also subscribed to the ethical position that "Biotic diversity has intrinsic value, irrespective of its instrumental or utilitarian value". He considered this to comprise a primary "normative" premise. Adopting such an ethical position would enable humans to emerge from the "dualistic" and "exploitative" environmental ethic to a more unitary perception (Soulé 1985, 731). He further argued that the inherent value existing in any given species is not something planted in it by humans, but is a product of its long evolutionary lineage. To this, we may also add its role – however insignificant – in the structural composition and functional properties of the ecosystems in which it resides, and which in turn is ultimately a part of the larger biospheric evolution. Just as our concept of humanism regards all humans – prosperous or poor, important or insignificant – as possessing an intrinsic value, it is now essential to extend that humanism beyond the confines of the human species.

In this regard, Michael W. Fox raised an important issue in the extrinsic-intrinsic value debate. He reasoned that extrinsic and intrinsic values may not be mutually exclusive. For example, a deer has high extrinsic value in terms of its importance to predators in the food chain, though at the same time it also has a life and an intrinsic value of its own. Fox also pointed out that while a limited number of predators cannot decimate a prey population, the high population load of human species acting as predators can have a devastating effect on the other organisms. He also contested the belief of some philosophers that the inherent value of domesticated animals, which have been designed primarily for human use, is less than that of wild animals. Fox attributed this attitude to the influence of an anthropocentric worldview. This same anthropocentric worldview that also has "racist and speciesist" implications tends to ascribe higher intrinsic value to a "being that is more intelligent and self-aware than another". This automatically leads to the fallacy of granting more intrinsic value to a sentient mammal or a human than to soil microorganisms, which nevertheless have no less – if not more – contribution to the functioning of the ecosystems which they inhabit and its "biotic community". He also reminds us of the "inherent potential of rocks, embodying trace minerals, and water". With such a realization, one could fully appreciate St. Francis calling "sister water", because while in a conventional sense, it lacks consciousness and sentience to be considered inferior to humans; in "her inherent potential" and "her extrinsic value to all life", she could be said to personify the "saintly virtue" of selfless service to all. Thus, in a figurative way or in terms of her extrinsic value, she shows the hollowness of the claim of "human superiority" (Fox 1989, 159). Finally, Fox asked humans to question themselves on their own extrinsic value toward the well-being of the other members of the biotic community as well as the ecosystems that they inhabit.

In the backdrop of such thoughts on the need for a changed perception of the values of the nonhuman world, Selma Bayram contended that only a proper ethic could change the social attitude toward nature and its conservation which is essential to resolve the environmental problems (Bayram 2012). The dominant anthropocentric ethics which postulates that only humans have intrinsic value, and the rest including nature only has extrinsic or instrumental value to serve human ends, is considered inadequate by many ethicists and environmentalists. In this context, a proper valuing of nature is very important because of the human tendency to value almost everything that surrounds them including other humans such as family and friends, food, the land, the houses that they live in, "political views, religion" (Bayram 2012, 1087), etc. Bayram referred to Rolston who wrote long ago that "an adequate ethics" for the various nonhuman communities on the Earth is lacking (Rolston 1988, xi) and humans are unable to value wild nature. Nature is more often viewed as "a means, not an end in itself" (Rolston 1988, 1), since it does not have any intrinsic value. Rolston challenged this long-held view that nature is bereft of any intrinsic value. He showed that numerous entities in nature have "objective intrinsic value", that is, values that are naturally (and objectively) possessed by that entity. For example, he cites the nutritional qualities of a potato, which developed there as a part of the plant's evolution, and not because humans exploited this property for their own nutrition and consequently ascribed an extrinsic value to it. Similarly, we may cite the presence of medicinal compounds in many plants, which did not evolve there to cater to human needs but are a product of the evolution of those plants. Therefore, these values are primary objective intrinsic values. When humans learned to use these compounds for medicinal purposes, they assigned a value to them, and termed these as extrinsic values. Therefore, if a natural entity has extrinsic value for humans, it does not mean that value is merely extrinsic. Its origin was as an intrinsic value, in many cases which the entity possessed – or acquired in the course of its evolution – long before the appearance of humans on the Earth.

1.5 Ecosystem Services versus Intrinsic Value

An ecosystem services (ES)-centered approach presently occupies a dominant place in the rationale for the conservation of nature. This has often relegated to the backseat – at least among a large number of ecologists, environmental scientists, planners, and policy makers – the view that nature should be protected and conserved because its components have intrinsic value.

Chelsea Batavia and Michael P. Nelson pointed out the prevalent tendency among a group of conservationists that utilitarian approaches like the ES work better than the more philosophical intrinsic value (IV) approach in influencing policy making, public support, and planning. The proponents of an ES approach assert that this approach will also protect those components of nature that may not be directly useful to humans. Thus, they do not question the "justification" or "appropriateness" of the intrinsic value approach, but reject it because it "does not work". In other words, they are simply treating intrinsic value as a strategy, and not as a moral or ethical standpoint. If this trend continues, then these ethicists are apprehensive that the ES approach may totally overshadow intrinsic value approach and push it to the backseat. However, they, along with some other ethicists, propose that concepts like ES which focus on the importance of human benefits from nature "should support, but not define" the basis of conservation (Batavia and Nelson 2017, 372).

Relying solely on ES approach is yet another short-term thinking to protect and save nature. Such approaches have the limited goal of producing a blueprint of action that will serve as a mitigatory, stop-gap measure for producing localized results. On the other hand, intrinsic value is a philosophy and aims for long-term and lasting changes in human perception of nature, which is expected to produce a paradigm shift in policy making and public opinion. It is true that the presence of intrinsic value in nonhumans may not have found unequivocal acceptance among all quarters of public policy makers and planners, who are still greatly influenced by anthropocentric approaches of which ES is a relatively recent one. Nevertheless, the deep philosophical roots of the intrinsic value concept cannot be replaced by other, shallower concepts such as instrumental, extrinsic, or utility values, and ecosystem services.

1.6 Relational Value

The instrumental value versus intrinsic value debate is carried further by Chan et al. (2016) when they argued that it is not enough to focus on either intrinsic or instrumental value to adequately reflect the "views on personal or collective well-being, or 'what is right', with regard to nature and the environment". Instead, they laid stress on "relational values", which include, among others, "preferences, principles, and virtues associated with relationships", and the Aristotelian "eudaimonic" values", that is those associated with "a good life" (Chan et al. 2016, 1462). They further pointed out that these values are not "present in things" but are derived from "relationships and responsibilities to them". They cited many examples from different western, eastern, and indigenous philosophies, social movements, and community practices and beliefs. In essence, they reverted to an anthropocentric approach to solve the environmental crisis, because all these values and relationships pertain to human perceptions of nature in the attempts to find the best way to use nature in human welfare. John J. Piccolo had, however, observed that relational values are a form of instrumental value. He further reiterated that contrary to the claim of Chan et al. (2016) that the intrinsic value concept has not been able to influence "the public and policymakers", its increasing recognition has contributed to the growing importance of conservation biology in the public sphere and economic planning (Piccolo 2017). Piccolo also referred to Muraca (2011), who constructed a "map of moral significance" that classifies instrumental value as a "subcategory of the wider class of 'relational values'". In her axiological classification scheme, values can be first categorized into "intrinsic values" or "moral inherent value" and "relational values". The latter, in turn, are of "fundamental-relational" or "functional-relational" types. The functional-relational category is further divisible into "intrinsic-eudaimonistic" and "merely instrumental" sub-categories. Thus, the debate cannot be between relational and instrumental values, because the latter according to Muraca (2011) is a subcategory of the former.

Going beyond these discussions about the finer points of categorization of values, it can be argued that relational values cannot play the same role as that played by intrinsic value in long-term conservation and environmental protection. Relational values include "preferences, principles, and virtues associated with relationships, both interpersonal and as articulated by policies and social norms", and are based on their "consistency with ... justice, care, virtue and reciprocity" (Chan et al. 2016, 1462–1463). It would be pertinent to point out here that all these preferences, principles, etc. may change with a shift in anthropogenic demands and priorities, and therefore, may fail to assure long-lasting protection. Further, we may also ask the question that toward whom would justice or care be directed? What is to ensure that humans will not prioritize justice and care to the members of their own species, which in turn may compromise human treatment of nonhumans?

Anne W. Rea and Wayne R. Munns tried to offer a reconciliatory course of action in the ongoing debate between the intrinsic versus instrumental value of nature. They suggested that considering both intrinsic and economic properties of nature could yield more meaningful outcomes than by simply applying economic valuations of nature. At the same time, they also pointed out the difficulty of quantifying the value of natural systems by way of monetary valuations (Rea and Munns 2017).

An attempt to reconcile the concepts of intrinsic and extrinsic/instrumental values and to reassess instrumental values was also made by Patrick Baard. He suggested that an entity can be said to have instrumental value if it serves as a means to another entity having intrinsic value. This raises the issue that if there are several other equally viable entities that can serve as means to the entity having intrinsic value, then all these entities would have equal instrumental value, and can be substituted by one another. Or there could be two situations: one in which several means in a chain are necessary to reach the end, but neither of these means is individually capable of reaching the end. In the second situation, an entity could be capable of serving as the sole mean to reach the end, that is, something of intrinsic value (Baard 2019).

In such a case, one could ask whether the entity being one of the means and the one individually capable of reaching the mean be regarded as having equal value. Again, if something is the only mean to reach an end, then would that entity and the entity that represents the end – that is the one having intrinsic value – be considered equally valuable? If there are five plants that contain five different chemicals each of which can cure the same disease in humans (the entity having an end or with intrinsic value), then can we allow four of them to get decimated or extinct because there is another that will serve their purpose? The next question is if there is only one plant that has a chemical that can cure the disease, then does the value of that plant be considered equal to the entity representing the end, that is the humans? If we decide our conservation policies based on such criteria, would these be justified or even effective in view of the biodiversity losses that we are experiencing?

Till the instrumental-intrinsic value debate is fully resolved, only humans are generally recognized as having intrinsic value, though this view is contested by biocentric and ecocentric thinkers. If the human uniqueness of possessing intrinsic value is accepted, then many inconspicuous and "lower" organisms will not have any instrumental value either, because these are not (at least directly) serving as means to humans. Microorganisms that contribute to the fertility of soil, and in turn enable the plants useful to humans to flourish, do not have any instrumental value because they are not means to an entity – humans – that has intrinsic value. However, Robin Attfield (Attfield 1999) argued that some entities such as ecosystems have such high instrumental value that they are considered more valuable than even the entities having intrinsic value – humans – because they facilitate their survival and well-being. Despite this importance, ecosystems cannot be ascribed intrinsic value. The value of biodiversity could also be determined in the same way because of its paramount importance to humanity. In this view, it is not necessary to ascribe intrinsic value to entities like ecosystems and biodiversity, because a recognition of their high instrumental values is adequate enough to protect them. However, opinions differ on this contentious issue.

1.7 Rolston's Views on the Values of Biodiversity

Before drawing this chapter to a close, I would like to make an extensive discussion of the meticulous and highly penetrative treatise of Holmes Rolston, III on the subject of values of and/or present in plants, animals, species, ecosystems, and the biosphere (Rolston 1994). Rolston examined the question of values in the context of some of the earlier observations that value is not an inherent property in an object, but only exists "in a relation to an appreciating mind" (Windelband 1921, 215, as cited in Rolston 1994, 13). When this mind is taken away, the value also disappears. Rolston also examined the views of Bryan Norton who said that it is futile to look "for a value in living things that is *independent* of human valuing" (Norton 1991, 251, as cited in Rolston 1994, 13). Rolston contended that besides valuing instrumentally, humans can also value nature intrinsically. Thus, value goes beyond the narrow confines of being "human-regarding" ("anthropocentric") to "human-generated" ("anthropogenic"). Rolston cited the example of visitors in Yosemite National Park in California who do not regard the giant sequoias there as a source of timber, but "as natural classics, for their age, strength, size, beauty, resilience and majesty". Nevertheless, this value originates from and is dependent on human viewing and valuing, but is not "human-regarding". It is, therefore, subjective, though it is "placed objectively in the tree" (Rolston 1994, 14). The same can be said about the rock strata of the Grand Canyon and their color bands, the stalactites of the Mammoth Cave of Kentucky, or the trilobite fossils. Rolston said that "the attributes under consideration are objectively there before humans come, but the attribution of value is subjective" (Rolston 1994, 15). However, the scenario changes when we observe the life of an animal. Rolston cited the example of a mother bat, who can use sonar to move through the maze of caves, catch insects, and suckle her young. She (and any other animal) not only has a good of her own, but she also can value the environment around her, her food, and her progeny. So, this shows that value can not only be non-anthropocentric, it could also be non-anthropogenic. Rolston concluded that "Animals maintain a valued self-identity as they cope through the world. Valuing is intrinsic to animal life" (Rolston 1994, 16). What about plants? Plants can grow, reproduce, mobilize water and nutrients, produce their food by photosynthesis, can attract pollinating insects and defend against harmful ones by secreting various chemicals or by developing thorns, and so on. Even then, many ethicists maintain that plants may have their own good, but they are unable to value, because they cannot "feel anything" (Rolston 1994, 17). As Peter Singer said, "if a being is not capable of suffering, or of experiencing enjoyment or happiness, there is nothing to be taken into account" (Singer 1976, 154). Robin Attfield made an interesting observation "even if trees have needs and a good of their own, they may still have no value of their own and may still be due no consideration in their own right". However, while examining the paradox that "trees have interests but no value of their own", he said "that trees can after all be of intrinsic value, even though we seldom need to take account of it in practice" (Attfield 1981, 35). He finally concluded that "living creatures in any case, it would seem, characteristically have a value of their own" (Attfield 1981, 53).

Today, we find that Rolston's conclusions are highly relevant in the context of changing definitions of cognition and sentience, which are likely to develop further with the expansion of our knowledge of plant and animal communication and neural and/or sensory perception and evolutionary strategies. For example, the use of volatile organic compounds (VOCs) by caterpillar-damaged plants to communicate with neighboring individuals has been known since the 1980s and has now received conclusive experimental support. A recent review of these findings (Meents and Mithöfer 2020) shows that many plants like sagebrush, blueberry, poplar, and others could communicate via VOCs to neighboring plants up to a distance of 50–60 cm or so. Root exudates also play a role in belowground plants to communicate in response to an aboveground stimulus (Elhakeem et al. 2018). Not only higher plants, fungi have also been shown to use electrical spiking activity to transmit information within mycelial (root-like structure of fungi comprising branched hyphae) networks. These spikes, arranged into clusters, could be said to comprise the "language" of fungi (Adamatzky 2022). These findings challenge the views of those ethicists who refuse to acknowledge intrinsic value in plants (a more detailed discussion on the issues of plant and animal feelings and consciousness has been made in Chapter 3).

1.8 Values in Entities at Higher Levels of Organization

Rolston also looked into the values of entities at higher levels of taxonomical and ecological organization, such as the species, the ecosystem, the Earth, and nature as a whole. Humans value many species of plants and animals, especially those that are endangered. Rolston raised the question whether a species itself (as a unit) has value, irrespective of human valuation. A cursory analysis may say that it does not, because it can neither enjoy nor suffer unlike an individual belonging to a certain species. However, Rolston argued that a species has an end or an ultimate aim or object, that is, a "telos". To explain this further, we can say that each species represents the gene pool of all the individuals belonging to that species. This assemblage of genes is perpetually evolving, with natural selection operating to select better structures and more appropriate strategies. Therefore, it can be said to have a telos, an end. Rolston further reasoned that when an individual wolf kills an elk, the elk suffers, but the elk species *Cervus canadensis* does not, because predation by wolf (and other predators, if any) helps to keep their numbers within sustainable levels. Thus, we can probably say that this is the way a species values itself and the other species with which it interacts. Rolston also turned his attention to ecosystems, which we know are complex and interacting assemblages of plants, animals, microorganisms, decomposing organic matter, and the inorganic domain comprising temperature, rainfall, humidity, and chemical properties of soil and water. Rolston further stated that ecosystems are "selective systems", which select for a number of system-level criteria such as diversity, stability, resilience, adaptive fitness, and both positive and negative interactions among species to enable the ecosystem to function more efficiently. All the functions of ecosystems are controlled by cybernetic loops and homoeostatic (self-regulatory) mechanisms. Because of these properties, Rolston argued that the terms intrinsic or instrumental are not appropriate to describe the mode of valuing of ecosystems, which, he said, have "systemic value" (Rolston 1994, 25). Rolston also took the exercise of valuing to the planetary level, taking the entire Earth under its purview. Is the Earth only instrumentally valuable to the life that it harbors? Anthropocentric reasoning would suggest that "Earth is no doubt precious as life support, but it is not precious in itself" (Rolston 1994, 27). Earth neither has the vitality of an organism, nor the capability to perpetuate like the genes, and is not a compact unit like an ecosystem. Rolston questioned this view and pointed out that there is a higher level of organization in the form of biosphere, and that Earth has definite planetary boundaries to indicate its existence as a finite entity. His view gets support from the science of ecology, which tells us that there is a continuity in the successive levels of organization beginning at the atomic and molecular level to proceed into higher levels of organization such as ecosystems, landscapes, biomes, and the biosphere, the latter interacting and integrating with the atmosphere (realm of air), the hydrosphere (realm of water), and the lithosphere (realm of rock or soil, that is, land). Therefore, Earth is not a mere rock pile, it is a macrosystem. This perception of Earth began with the concept of biosphere put forward by Eduard Suess in 1875 (Billings 2020), which was further elaborated by V.I. Vernadsky (Vernadsky 1945; Cancela da Fonseca 2000). Hence, if Rolston found systemic value in ecosystems, then Earth deserves to be given a "macrosystemic value". Further, the species during its evolution selects certain traits, while rejecting others. Similarly, the ecosystem and the biosphere also select certain "systemic values" over certain other values. Hence, these systems can also be termed as "valuers", and not merely the subject of valuation.

1.9 Conclusions

The discussion on the concept of value calls for fresh thinking and appraisal. The "old paradigm" says that there cannot be any valuing without a valuer, a role exclusively reserved for humans. At the most, this capability could be extended to include sentient animals. Other entities like plants, lower animals, ecosystems, Earth, and nature cannot be treated as "bona fide valuers" (Rolston 1994, 29). With a vast amount of new knowledge accumulating on cognition and communication in microorganisms, plants, and both lower and higher animals, and the nature of self-regulatory mechanisms and evolutionary strategies of ecosystems and the entire biosphere, perhaps it is time that this old paradigm is replaced by a new one. The brilliance of Rolston lies in the fact that he could give us a glimpse of this emerging, new understanding three decades ago. It may also be noted here that the concept of intrinsic value and the need for its recognition in nonhumans by humans as a species is soundly ensconced in certain strands of western environmental philosophy and movements such as Deep Ecology and others. It is a major trend in eastern philosophical thoughts such as Hinduism, Buddhism, Jainism, Dao, Shinto, and numerous folk practices and nature religions all over the globe. It has also permeated the writings of a large number of poets, novelists, and writers of various other genre of works. A detailed discussion of these thoughts is made in Chapters 5, 6, and 7 of this book. While admitting that acceptance of the concept of intrinsic value in nonhumans may not have been unequivocally accepted in all quarters of public policy makers and planners, who are still greatly influenced by anthropocentric approaches of which ES is a relatively recent one, the deep philosophical roots of the intrinsic value concept cannot be replaced by other, shallower concepts such as instrumental, extrinsic, or relational values.

2

ETHICAL FRAMEWORKS

Anthropocentrism, Biocentrism, and Ecocentrism

2.1 Introduction

In order to study the postulates of ecocentrism, we need to understand the other ethical positions in environmental ethics, and the basic distinctions between these and ecocentrism. The readers can also refer to several exhaustive works (e.g., Rolston 1988, 1994; Sessions 1995; and others). Anthropocentrism (Greek – Anthrōpos: human; Kéntron: center) is an ethical position that recognizes intrinsic value in humans alone. All other entities – both living and non-living – have only extrinsic or instrumental value by virtue of their usefulness to humans. In other words, only humans represent ends in themselves, while all the "other things ... are means to human ends". Humans occupy a pivotal position because only they have the "cognitive ability to formulate and recognize moral value". Opposed to anthropocentrism are the ethical positions maintained in biocentrism (Greek – Bios: life; Kéntron: center) and ecocentrism (Greek – Oikos: House; Kéntron: center). As their names indicate, biocentrism recognizes intrinsic value in living organisms, and ecocentrism recognizes it in living organisms and the ecosystems in which they live.

2.2 Postulates of Anthropocentrism and Their Critiques

Two categories of anthropocentrism – "narrow" and "enlightened", or "weak", or "broad anthropocentrism" are usually recognized (Goralnik and Nelson 2012, 145). John Passmore advocated for a strong anthropocentrism where only humans have rights, and are viewed in isolation from the environment. Other anthropocentric ethicists like Kristen Schader-Frechette

DOI: 10.4324/9781003481058-2

and William Frankena also dismissed the need to develop a "more inclusive ethic" (Goralnik and Nelson 2012, 146). In contrast to this position, weak or broad anthropocentrism recognizes the "aesthetic, educative, or restorative" values of the environment other than mere instrumental value. Bryan Norton suggested that there is no need to develop a radical environmental ethic that awards intrinsic value to nonhuman entities. Instead, human exploitation of the environment needs to be regulated in order to maintain a harmonious existence with nature. Thus, this form of anthropocentrism recognizes the "ecological relationships" of humans (Goralnik and Nelson 2012, 149). However, it is apprehended that if humans recognize only instrumental value in nature, then in the event of any conflict between human and nature's interests, the former is most likely to receive preference over the latter. For instance, if a road is to be constructed through a forest to provide improved communication to human communities in that area, then the utilitarian principle of "greatest good for the greatest number" is most likely to prevail to dissect the forest into smaller fragments, which may not be able to support the ecological needs of macrofauna such as elephants. This may cause their extinction from that area, or lead to human-elephant conflict. On the other hand, if intrinsic value is recognized in elephants, and still better, in the entire forest ecosystem, then people are likely to go for a more benign alternative. In such instances, anthropocentrism fails to provide the ideal ethical-philosophical support to human actions. However, Norton (1984) reasoned that the "rights or interests" of neither nonhuman species nor future humans could serve as the basis for deriving a valid environmental ethic. Instead, he argued that it is not at all necessary to frame a distinct environmental ethic other than an anthropocentric one. He asserted that "standard ethics", which take care of the rights, interests, and duties of humans, are sufficient to enable judicious environmental policies and actions. This is because nonhuman entities have no separate value outside those of human values. Consequently, an anthropocentric environmental ethic is adequate for environmental protection, and there is no necessity for an anthropocentrism versus non-anthropocentrism debate. Norton also distinguished between "strong" and "weak" anthropocentrism on the basis of fulfillment of "felt" and "considered" preferences, respectively. Felt preferences simply reflect the demands of humans, while in considered preferences the demands are in conformity with a worldview that has been adopted rationally on the basis of scientific findings and a supporting metaphysical framework. Therefore, strong anthropocentrism considers human preferences and demands indiscriminately, and does not question or disapprove values or practices that harmfully exploit nature. Weak anthropocentrism, on the other hand, distinguishes between rational (and legitimate) and exploitative values and practices and applies a sort of screen or filter. Consequently, it propagates an ethical or moral value system that advocates a harmonious existence of humans with nonhuman nature. Weak anthro-pocentrism is able to provide a framework for developing an ethical system for the welfare of both humans and nonhumans. It attempts to maintain a balance between the interests of the two groups. Because of the considerable mitigating effects of a weak anthropocentric ethic, the question of whether nature has intrinsic value or only instrumental value loses its relevance. Norton further cited the examples of Hinduism and Jainism, which prohibit the killing of certain animals including insects for their own spiritual uplift-ing rather than protecting these life forms *per se*. He also gave the exam-ple of Henry David Thoreau, who, according to Norton, did not ascribe intrinsic value to nature, but believed that humans can improve themselves spiritually by realizing the spiritual treasures contained by nature. Similarly, humans could subscribe to a worldview that uses scientific reasoning for a harmonious co-existence with nature in light of the evolutionary concepts of Darwin. Thus, there is no need to recognize intrinsic value in nature in order to protect it from human excesses. In a critique of the statements of Norton, it is necessary to mention here that both Hindu and Jain worldviews sponta-neously recognize intrinsic values in nature, and advance beliefs in the unity of all life. However, these worldviews also state that prohibition of killing of animals or harvesting of plants can bring spiritual uplifting to those who observe these rules. Thus, these two sets of beliefs do not negate each other, and uphold two faces of the same ethic. Coming to Thoreau, he wrote in *Walden* that "Sympathy with the fluttering alder and poplar leaves almost takes away my breath" (Thoreau 1995, 71). In these words, we can detect his identification with these nonhuman lives, which can be said to reflect his recognition of something beyond their instrumental value. And in his essay *Walking*, he wished "to speak a word for nature, for absolute freedom and wildness, ... to regard man as an inhabitant, or a part and parcel of Nature" (Thoreau 1862). These feelings indicate that it may be wrong to conclude that Thoreau did not recognize intrinsic values in nature (see Chapter 7 for more on Throreau's worldview).

Tim Hayward had also defended anthropocentrism, which he dubbed as an "ontological error", but justifiable from an ethical viewpoint. He claimed that condemning anthropocentrism as a whole may adversely affect some of the "legitimate human concerns". In his selective appraisal of anthropocen-trism, he pointed out a paradox, where developments in modern science have made it clear that humans are not at the center of the Earth or the universe, but a product of evolution like the other forms of life. This realization shows the wrong ontology of anthropocentrism. On the other hand, modern sci-ence also shows that humans stand "apart from the rest of nature" in many ways (Hayward 1997, 50). Since it cannot be wrong for humans to have a particular concern for the "legitimate interests" of fellow humans, anthro-pocentrism *per se* cannot be an approach to be criticized. Instead, Hayward blames speciesism, which promotes illegitimate interests of the human species at the expense of other species, and human chauvinism, which fails to respect other species because they lack certain chosen human endowments such as "rationality, language, subjectivity etc." Speciesism and human chauvinism taint anthropocentrism with unethical approaches and properties and need to be removed. However, though it may be possible to eliminate human chauvinism, total removal of speciesism would be more difficult. In any case, abolishing anthropocentrism in totality is untenable for several reasons. Firstly, even to rectify speciesism by according moral status to nonhumans, one has to use a human benchmark. Secondly, the values to be recognized in nonhumans would be done by a human. Consequently, applying human standards cannot be avoided. Thirdly, human "self-love", and love and concern for the other members of the human species, could form the basis for loving the nonhumans too. Therefore, anthropocentrism need not be criticized as long as it is concerned with legitimate human needs and interests. Further, when other species or the environment are affected, the harm is not caused by humanity as a whole, but by certain interested group/s. For example, hunting a species to extinction or pushing it to the brink of extinction is done by a small group of humans such as poachers, and deforestation is done by those deriving economic benefits from this act. Therefore, it is not correct to blame the entire humanity for these acts perpetrated by a section of the human race. Hayward also contended that while biocentrism can be accommodated, ecocentrism is untenable, because no environmental harm could be conceived without referring to effects on one species or the other. Further, the role of ecosystems is to serve as "life-support" systems for humans or other species. Besides serving as habitats for various species, and for satisfying human needs such as "spirituality or beauty" (Hayward 1997, 61), there is no justification towards their conservation for their own sake. Hayward, therefore, dubbed ecocentrism as "radically indeterminate" and unable to serve as a point for launching anti-anthropocentric arguments. Two counterpoints to Hayward's arguments can be raised here. Firstly, not only poachers but numerous noblemen and other trophy hunters, "sportsmen", and other hunters are responsible for decimating the world's wildlife, while a large number of factors such as urban, industrial, and agricultural expansion have reduced biodiversity through encroachment of wild habitats. Therefore, humans cannot escape their responsibility as a species in degrading the environment by shifting the blame onto the shoulders of poachers and timber loggers. It is from this realization that environmental ethics has focused on the question of giving moral considerations to nonhumans, and has generally rejected the anthropocentric basis of awarding values to nature. On the other hand, most practical measures to address environmental problems are based on anthropocentric approaches, thereby creating a serious contradiction. Some frequently debated aspects are: whether only nonhuman animals or some selected species of plants and animals or entire ecosystems should be the subject of moral considerations; and whether this should be confined to only natural ecosystems or man-made ones as well. Andrew Light suggests an alternative path that environmental ethics could follow to play more meaningful roles as a "public philosophy" (Light 2002) for resolving environmental problems. In this path, an environmental ethic need not summarily reject the anthropocentric logic of environmental conservation, because an environmental ethic that totally rejects anthropocentric values is inconsonant with the other forms of environmental inquiry such as environmental sociology, environmental economics, environmental health, environmental law, etc., which operate primarily in a human context. Therefore, an environmental ethic or philosophy totally ignoring anthropocentric arguments gets reduced to the status of a value theory, and fails to emerge as a powerful public philosophy. Light argued that this happens because the majority of people think that the most overbearing reason for environmental protection is to protect the interests and welfare of the future generations, as revealed by several surveys undertaken in the USA. Therefore, an anthropocentric environmental ethic is most likely to be able to influence the policy makers, though it is also not desirable for the anthropocentric ethic to be too "strong" to "reduce the value of nature to a crude resource instrumentalism" (Light 2002). He concluded that by adopting such an approach, environmental philosophy and ethics would assume a position of "methodological environmental pragmatism". Taking forward the argument of Tim Hayward (Hayward 1997) that not the entire humanity but certain interest groups are responsible for environmental degradation and destruction, David Kidner is of the opinion that the term anthropocentrism does not adequately describe the ideological basis of environmental destruction by human action, because the system that causes environmental damage affects both human and nonhuman well-being. Kidner questions whether the "human-centeredness" takes into account the interests of all humans or of a small section, and whether it only takes care of the short-term or long-term interests of the human race. In reality, the "human" around which anthropocentrism is centered represents the *"industrialized human"* (Kidner 2014). Research and innovations in science and technology are to a large extent dictated by the priorities set by the market, due to which the monetary value of an entity starts superseding or even pushing into the background its cultural, religious, ecological, or other values. The monetary orientation turns everything nonhuman into resources to be put to one use or the other. The status of nonhuman life changes from cohabitants and partners to biological resources. Ecosystems, especially their biotic and abiotic constituents, become natural resources and "raw materials", and even people turn into "human resources" (Kidner 2014). In a sense, this "industrocentric" ethic also robs humans – or at least certain classes of humans

– of their intrinsic value. As this process continues and intensifies, not only are the connections/relations that exist among the individuals in a community weakened or even eliminated, but the age-old relationships between the human and the nonhuman world are also eroded and replaced by an exploitative attitude. Thus, what is termed as "human interests" in a more generalized manner, in reality, represent "*industrial* interests". Hence, holding anthropocentrism responsible for the ecological ills would prevent us from understanding the root cause of and identifying the agents responsible for ecological degradation and destruction. It is to be appreciated that the dominance of such an ethic not only reduces bio- and geodiversity, but also smothers cultural uniqueness and prevents the fostering of cultural values that promote the formation of bonds between the humans and the "others" in the biosphere (Kidner 2014).

Nevertheless, whether industrocentrism or technocentrism is specifically responsible for environmental degradation, industries or technologies ultimately cater to the ever-growing human demands and consumerism. Therefore, it stands justified to club all these together under the broad head of anthropocentrism, and the human species cannot be absolved of its responsibilities to the environment including its beleaguered biodiversity.

Helen Kopnina and her associates provided counterarguments against the justification of anthropocentrism given by Hayward (1997). Hayward tried to distinguish between legitimate and illegitimate human concerns and defend anthropocentrism on the assumption that it only defended the former. Hayward also revised the implication of the term anthropocentrism to denote the properties of compassion and humane attitude towards fellow humans. However, he did not clarify whether anthropocentrism also "takes into account the legitimate concerns for nonhuman welfare". This also does not clarify the nature of value present in nonhumans. Since anthropocentrism means "human-centered", it automatically implies that nonhumans do not have moral value. Regarding the argument that "self-love" can lead to love for others, it could also lead to taking care of the concerns of the human species at the expense of the interests of others. Hayward's contention that ecosystems should be conserved only because they provide "life-support system for humans" (Hayward 1997, 60) is contrary to ecocentric principles (Kopnina et al. 2018). Kopnina (2020) further pointed out that anthropocentrism does not have concern for nonhuman species that are not economically useful to humans. It is also not concerned with animal welfare.

2.3 Anthropocentrism versus Biocentrism and Ecocentrism

Despite the defenses and justifications of anthropocentrism – irrespective of whether we adopt it in its "weak", "enlightened", or "broad" form, or try to narrow down the erring component to industrial humans – many environmental ethicists have found anthropocentrism to be inadequate on many counts including its philosophical basis, and have consequently rejected it. As said earlier, two alternative ethical models that have been proposed include the biocentric (life-centered) and the ecocentric (ecosystem/environment-centered) ethics. Though named differently and differing somewhat in their scope, these two ethics essentially uphold similar principles, because organisms can rarely be viewed, judged, or treated outside their home or habitat, and any appreciation of an ecosystem or a habitat or respecting its integrity would remain incomplete without its resident life forms. Can we imagine a whale without the ocean or vice versa?

As has also been said earlier, anthropocentric approaches towards environmental protection including biodiversity conservation are considered more acceptable and realistic to policy framers and decision-makers rather than bio- or ecocentric visualizations. Burbery (2012) is of the opinion that biocentrism and ecocentrism aim to achieve "noble goals", though it is difficult for humans to emerge from their anthropocentric core and accept the position that all species are equal. That is why, it is more realistic to talk of weak anthropocentrism and opt for a steward position to replace the role of tyrant or master for humanity.

This compromise has almost always taken place in the history of modern conservation. We can recall the conflicts between the "conservationism" of Gifford Pinchot and the "preservationism" of John Muir in this context.

2.4 John Muir and Gifford Pinchot: Proponents of Ecocentrism and Anthropocentrism

John Muir and Gifford Pinchot were two pioneers in the American conservation movement, though they came from highly disparate backgrounds and held distinctly different approaches towards conservation. Muir had a distinctly biocentric/ecocentric bent of mind, which is evident in many of his writings. For example, in his book *Our National Parks* (Muir 1901), he spoke about the limbs of Douglas spruce, cedar, fir, and hemlock "forming a forest kingdom ... in which limb meets limb, touching and overlapping in bright, lively, triumphant exuberance". In another place, he writes about the tall ferns fringing the rocks of the waterfalls to be soaked in a light spray, with the trees "leaning over ... like eager listeners anxious to catch every tone of the restless waters". In his mind's eye, National Parks were "places for rest, inspiration, and prayers" (Muir 1901, 20, 23, 30). Gifford Pinchot – on the other hand – was a highly competent forester who regarded the scientific farming of trees as the main objective of the science of forestry. Nevertheless, he also advocated for a prudent and sustainable use of the forest resources. Pinchot attached great importance to "people and natural resources" (Pinchot 1947, 325, as quoted in Meyer 1997), and had his own version of the utilitarian view, which asserted that "Conservation means the greatest good to the greatest number for the longest time" (Pinchot 1910, 48, as quoted in Meyer 1997). Smith (1998) cited a statement each of Muir and Pinchot, which perhaps reflects their attitudes towards nature. Muir appealed to save the towering redwoods by saying that "while fire and the axe still threaten destruction, make haste to come to the help of these trees, our country's pride and glory" (Muir 1903, as quoted in Smith 1998). Pinchot's approach was decidedly anthropocentric and oriented towards use value, when he said that "It is almost impossible to bring home to the average man the economic importance of this great national resource. But without cheap lumber our industrial development would have been seriously retarded" (Pinchot 1901, as quoted in Smith 1998). These two utterances clearly show the differences in perception and priorities of these two stalwarts of the American conservation-preservation movement. Pinchot's arguments found favor with President Theodore Roosevelt (Meyer 1997) and consequently prevailed in the great debate that he had with Muir on the construction of a dam in the Hetch-Hetchy Valley of the Yosemite National Park to supply the city of San Francisco with water. Muir and the preservationists were totally against the idea of human interventions in the Hetch-Hetchy and the Yosemite. However, Pinchot's utilitarian logic attracted more popular attention, and President Woodrow Wilson signed the bill, followed by an Act to make the dam a reality (Smith 1998).

2.5 Dominance of Anthropocentrism

It is also noteworthy that most international reports, conferences, conventions, and other documents, e.g., the United Nations Conference on Human Environment (UNCHE) in Stockholm, 1972, the World Conservation Strategy (IUCN 1980), the World Commission on Environment and Development or the Brundtland Report (WCED 1987), the United Nations Conference on Environment and Development (UNCED) in 1992, and others, have largely advanced anthropocentric arguments for the conservation of biodiversity, emphasizing the essential role that biodiversity played in human welfare and survival. The concept of ecosystem services emphasizes the role of ecosystems in enabling the human societies to survive and thrive. The Intergovernmental Science-Policy Platform on Biodiversity and Ecosystem Services (IPBES) assessment (IPBES 2019) highlights "nature's contribution to people" (NCP). The utility of ecosystem services was earlier promoted in the Millennium Ecosystem Assessment of 2005. Admittedly, some eco-centric views have also found their place in these documents. The UNCHE mentions the threats to wildlife, the WCED also prioritizes biodiversity conservation, the Convention on Biological Diversity speaks of the presence of intrinsic values of biodiversity, and the IPBES assessment acknowledges that sections of people, especially many indigenous communities the world over, find intrinsic value in nonhuman organisms, entire ecosystems such as rivers, lakes, forests, etc. and nature as a whole. Nevertheless, despite these mentions, the primary thrust in all the aforesaid initiatives is decidedly anthropocentric (Taylor et al. 2020). Making a case for ecocentric value perceptions of nature, Taylor et al. (2020) argued that such beliefs and perceptions are not only to be encountered in indigenous societies and other cultures, but are also to be found in the writings of naturalists like Charles Darwin and Alexander von Humboldt and environmental thinkers such as John Muir, Aldo Leopold, Arne Naess, J. Baird Callicott, Holmes Rolston III, and many others. They also cited the biophilia hypothesis (Kellert and Wilson 1993), which theorizes that the aesthetic appreciation for pristine ecosystems is an evolutionary legacy of the human race. Many scientists, bureaucrats, and policy-makers are of the opinion that since there is sufficient awareness among people about the dangers of the environmental crisis to human civilization, an anthropocentric value system should be enough to ensure adequate policy measures for amelioration. However, if we look at the present scenario, despite sporadic and case-specific success in saving species from extinction, biodiversity loss continues unabated. Hence, the anthropocentric rationale does not seem to be effective and needs to be supplemented with ecocentric recognition of intrinsic values in nonhuman entities. As a first step towards this paradigm shift, it is necessary to incorporate ecocentric values into the recommendations of the Convention on Biological Diversity, which holds its meetings of the parties from time to time.

If we take the debate on whether the value of biodiversity is only "instrumental or utilitarian", or "intrinsic or inherent" further, J. Baird Callicott suggested that because of the debated nature of the recognition of intrinsic value in nonhuman entities, many conservation biologists have put forward a purely utilitarian value of biodiversity, wherein biodiversity has a value merely "as a means to human ends". In other words, the prevalent view is anthropocentric, where only humans are qualified to have intrinsic value. The values of biodiversity are many, including that as goods such as food, fuel, fodder, and medicine; and as vital services through pollination, cycling of nutrients, fixation of atmospheric nitrogen, and cybernetic control. Further, there is value in information contained in genes, and their aesthetic, spiritual, or religious value, though the latter may also be regarded as intrinsic value. There is also the question of whether intrinsic value in biodiversity could be attributed "objectively" or "subjectively" (Callicott 2006, 111–113). As a response to these pro-anthropocentric reasonings, it can be argued that organisms are self-organized with homoeostatic (self-regulatory) properties, can control and direct their own actions unlike a machine, and have their own "goal and purpose" such as to reproduce, get food, grow, and multiply. Consequently, they could be said to have intrinsic value in an objective manner. All organisms have interests: for example, plants have interest in soil, nutrients, water, sunlight, etc.; animals have interest in sufficient food, mate, and shelter. However, intrinsic and extrinsic values are not "mutually exclusive", and an entity may possess both (discussed in the case of indigenous societies in Chapter 5). One advantage of conferring intrinsic value to biodiversity is that in that case sufficient justification has to be given before putting them to risk, just as is done in the case of humans.

2.6 Biocentrism and Ecocentrism

Talking about the alternative ethical approaches to anthropocentrism, we come across two major concepts: biocentrism (Gk: Bios: life; Kéntron: center) and ecocentrism (Gk: Oikos: house; Kéntron: center).

2.6.1 The Early Bio- and Ecocentrists

2.6.1.1 Albert Schweitzer

Though the elements of biocentrism can be found in many religions and indigenous worldviews (see Chapters 5 and 6), Albert Schweitzer (1875–1965) is one of the modern proponents of biocentric ideals. His life and teachings reflect his deep "reverence for life". In his book *Strassburger Predigten* published in 1966, which has been translated into English by Reginald H. Fuller as *Reverence for Life* (Schweitzer 1969), we can get a feel of his convictions. In a preaching delivered in 1919, Schweitzer spoke about the supreme importance of values like "empathy, compassion, sympathy" for "everything that is called life", which could also be expressed in a negative form as "Thou shalt not kill". Schweitzer emphasized the need for "love for all creatures, reverence for all being, compassion with all life, however dissimilar to our own". He went further to say that the "beginning and foundation of morality" stems from the reverence and compassion for all life. Without this foundation, morality remains merely "superficial" and cannot stand the tests that it has to confront now and then under diverse challenging situations. The detailed knowledge of the scientist and the perceptions of a farmer of the blossoming buds in spring are both ultimately confronted with the "riddle of life" and its enigmatic nature (Schweitzer 1969, 114–116). However, while extolling the value of empathy and compassion, Schweitzer did not recognize these values in nature, which to him appeared to be producing the diverse creatures "in the most meaningful way" and destroying them "in the most meaningless way". Nonhuman organisms did not have the ability to feel compassion for creatures other than their own offspring, and killed the other forms of life. Only humans were privileged to see the light of compassion, and could transcend "the ignorance in which the rest of creation pines". The appearance of this ability of humans was the "great event in the evolution of life", which brought "truth and goodness" into the world (Schweitzer 1969, 120–122). In this sense, Schweitzer's biocentrism can be said to be permeated with a sort of "soft" or "weak" anthropocentrism. Humans assume the role of a benign and compassionate steward in the vision of Schweitzer.

2.6.1.2 Aldo Leopold

A couple of decades later, Aldo Leopold breathed fresh life into ecocentric thoughts in his own inimitable way. In his *A Sand County Almanac and Sketches Here and There* (Leopold 1949), he regarded "man" to be "only a member of a biotic team". Leopold raised questions about the oft-proclaimed "love for and obligation to the land of the free ..." and exposed its anthropocentric basis and lack of inclusivity. This was because the human "love for land" neither included the soil, resulting in soil erosion; nor did it care for the water that was being grossly polluted and diverted for human use. Human actions also led to the decimation and even extermination of plants, animals, and all other forms of nonhuman life. However, in order to lead an ethical existence, an individual has to be a part of a community, where its role is not only to compete but also to cooperate with the other members of the community. Based on these premises, Leopold proposed his "land ethic" that visualized an image of the community including not only the humans but also "soils, water, plants, and animals, or collectively: the land". Adopting a land ethic also transforms the role of the human species from that of a "conqueror of the land-community to plain member and citizen of it". This means that humans not only respect the other members of the community, but have the same feeling for the community as a whole. Leopold also reminded us that the outcome of the conqueror role in human history has almost always been "self-defeating", primarily because of the large gaps in the knowledge of the ruler about the values and functions of the community. He, therefore, wished for "the concept of land as a community" to penetrate "our intellectual life" and shape our "ecological conscience" (Leopold 1949, 204–205, 207). At the root of Leopold's arriving at such conclusions in the 1940s – when perhaps not even a shadow of the environmental crisis of the future could be perceived at the horizon – was his years of following nature with both its myriad creatures and non-living entities, which enabled him to imbibe deep within him the ethical recognition of the nonhuman elements of the land community. In his writings, one comes across the bewildered meadow mouse that is baffled as its elaborate tunnels through grass stumps covered with snow suddenly melt away in the January thaw, and expose it to the danger of being picked up by the rough-legged hawk. The synchrony of their lives with the periodic oscillations of nature is succinctly expressed by Leopold as "To the mouse, snow means freedom from want and fear." On the other hand, the hawk "is well aware that snow melts in order that hawks may again catch mice..... to him a thaw means freedom from want and fear". The arrival of the geese at the marsh signals the advent of spring, when these visitors chat with the sandbars and greet "each newly melted puddle and pool". At such moments, the author of the *Almanac* wished that he were a muskrat, half-submerged in the marsh. But the life of the geese is not without its pathos, with the ruthless human hunters relentless after their flesh in the winter. To the chagrin of Leopold, he found that geese mostly lived in families or groups of six, and the lone geese, which were often seen flying around and honking with a "disconsolate" tone, were the survivors of their families exterminated in the winter shooting spree, "searching in vain for their kin". Beyond the more spectacular and visible wolf, deer, muskrat, and geese, Leopold's ecocentric gaze also fell upon the nondescript *Draba* flower, which also heralds the arrival of spring, appearing almost everywhere in "small blooms". It is hardly noticed by anyone, not given much importance by botanists, "nobody eats it" and "no poets sing of it" [though in Chapter 7 we will find poets expressing their love and respect for the "humble" in nature, irrespective of their size, beauty, and prominence] ... "it is no spring flower, but only a postscript to a hope". As opposed to this frail herb, the prairie has the "bur oak", which has withstood the annual prairie fire by growing "bark too thick to scorch" (Leopold 1949, 4, 19, 22, 26–27). A life spent wandering through, studying, and interacting with nature led to Leopold's spontaneous recognition of its intrinsic value. This realization enabled him to say that "Only the mountain has lived long enough to listen objectively to the howl of a wolf". And he was never hesitant to admit his mistaken notions; when as a young trigger-happy forester, he shot dead a mother wolf in front of her cubs in the belief that fewer wolves would mean more deer and better hunting. To his utter dismay, he was to find later that the "wolfless" forests soon became the victims of overbrowsing by deer, finally leading to their "death". And then he understood why the wolf and the mountain did not subscribe to his idea that a mountain or forest without wolves meant more deer – thus becoming a "hunters' paradise" (Leopold 1949, 129). He realized that not only did the deer remained constantly afraid of being preyed upon by the wolves, but the mountain was also similarly afraid of the deer. And the worst part was that, while a deer population depleted by wolf predation could recover in a few years' time, it might take decades for a forest affected by deer overbrowsing to bounce back to its original verdure. This is an example of Leopold's ecocentrism, which, however, is never detached from the needs and demands of the society and its long-term well-being. He admonished the cowmen who made their ranges free of wolves, but did not limit the size of their herds. These ranchers did not "think like the mountain", leading to environmental problems such as that of the creation of "dustbowl" and erosion and sedimentation of rivers (Leopold 1949, 132). Human thinking often fails to see the "larger picture" and future consequences when they embark upon large-scale "doctoring"/tempering of natural systems, such as exterminating wolves from a forested landscape or from a cattle range, or introducing chemicals without first ascertaining their long-term effects on the environment and organisms, or damming a river. In Leopold's words, they do not "think like the mountain".

2.6.2 Arguing for Biocentrism and Ecocentrism

2.6.2.1 Sentience versus the Other Conditions for Giving Moral Consideration

Coming back to the "not yet resolved" debate on whether moral status should be awarded to only the humans, or all sentient animals, or all plants and animals (biocentrism), or also to extend to include inanimate entities like ecosystems, larger landscapes, or the biosphere as a whole (ecocentrism) – we can take a look at some of the major arguments in favor of biocentrism and ecocentrism.

The basic position of biocentrism is that it does not accept the central position of humans in ethics and instead visualizes moral standing in all life forms. A biocentric ethic argues that all living organisms have their own good, and hence have intrinsic value. However, ethicists such as Peter Singer cannot strictly be termed as "biocentrists" since he only considered sentience (the capacity to have positive and negative experiences such as pleasure and pain) as the threshold of recognizing moral standing in an organism. Such a benchmark tends to exclude many lower animals and all plants (Attfield 2009; Animal Ethics 2021).

Again, it is Aldo Leopold who seems to have set the tone for a biocentric ethic when he wrote "A thing is right when it tends to preserve the integrity, stability, and beauty of the biotic community. It is wrong when it tends otherwise" (Leopold 1949, 224–225). As if to pick up the cue from where Leopold left it, Kenneth Goodpaster began by quoting these famous lines of Leopold in his article "On being morally considerable" (Goodpaster 1978). In this article, Goodpaster tried to answer the much-debated and discussed, yet not fully resolved, question, which simply put would be the "requirements for 'having standing' in the moral sphere" (Goodpaster 1978, 309). Goodpaster tackled this question in the context of endangered species and the treatment of animals by the human species as well as that of other contemporary issues. He did not think that rationality and sentience were essential for according moral standing of any entity. As said earlier, Peter Singer insisted on sentience – which he defined as "the capacity to suffer or experience enjoyment or happiness" – as the determining factor for giving moral status to any entity (Singer 1974, 108). Several other ethicists such as G.J. Warnock and William K. Frankena (Warnock 1971, and Frankena 1973: as cited in Goodpaster 1978) have also argued for sentience as a necessary criterion to give moral standing, and reasoned that if a being is alive but shows no evidence of the ability to feel pleasure and pain (for example trees), then they do not qualify to have any moral consideration (Goodpaster 1978). However, Goodpaster went beyond this limit and reasoned that "beings who have (or can have) interests can deserve moral consideration" (Goodpaster 1978). Joel Feinberg admitted that plants have biological properties like growth and can experience conditions which are beneficial or harmful to them. Therefore, plants "are capable of having a 'good'" (Feinberg 1974, as cited in Goodpaster 1978). Consequently, plants can have their own interests (e.g., growth, reproduction, etc.) independent of and distinct from the interests of humans, whose sole purpose may be to either nurture them till their products can be harvested or see them weeded out or felled. So, human interest turns a tree into timber, and killing it is termed as timberlogging. It is most likely that the tree neither wants to turn into timber, nor wants to be logged. When its old branches fall on the forest floor or it dies, its body serves as food and shelter for numerous species of animals and microorganisms. The question of giving moral standing to the latter is also a part of the ethical debate. Regarding the "epistemological problems about imputing interests, benefits, harms, etc. to nonsentient beings" (Goodpaster 1978), Christopher Stone pointed out that in the world of law, not only insentient but also inanimate entities such as "trusts, corporations, joint ventures, municipalities, Subchapter R partnerships, and nation states", "ships" (which are referred to in feminine gender), etc. are recognized as "right-holders" since long, though the jurists were initially highly skeptical of such a notion (Stone 1972). Goodpaster quoted Stone, who said that nonhuman, nonsentient entities could express their needs in a reasonably "unambiguous" manner. For example, he could judge with certainty when his lawn needed water through its dry and wilted appearance as well as other symptoms. Therefore, a lawyer could plead on behalf of a "smog-endangered stand of pines" to take appropriate and adequate control measures (Stone 1972). Thus, rights have been regularly extended to new entities in the history of law, making the rights-net progressively wider and inclusive. Further, because it is common practice to take decisions on behalf of entities whose interests are not expressed so explicitly, there should not be any objection to humans pleading or acting on behalf of living organisms whose wants are much more clearly expressed than, say, corporations, ships, or even nation states. Based on these contentions, Goodpaster (1978) surmised that "*being alive*" is a "plausible and nonarbitrary criterion" for granting moral consideration, which if accepted could lead to extension of this acceptance to the "biosystem itself". Such a paradigm shift is likely to have deep implications for environmental ethics and also towards the ability to take bold long-term actions to protect the environment. Robin Attfield deliberated on the question of whether trees have their own good, or are "good only for satisfying human interests" – the latter viewpoint held by many ethicists and philosophers (Attfield 1981). He traced this view to that of Thomas Aquinas (1225–1274 CE) and then to that of 20th-century thinkers like the British philosopher R.M. Hare (1919–2002) and the American philosopher Joel Feinberg (1926–2004). Attfield refuted these arguments and went on to show that trees could have a "good" because they have properties like growth, photosynthesis, respiration, flowering, and fruiting. At the same time, they are harmed by adverse factors like fire, flood, and landslip, and experienced decay and death. They could also be protected against human activities like logging and clear-felling. Thus, they share a number of properties with humans and animals, though it can be said that they "lack activity and self-motion ... beliefs, desires and feelings" (Attfield 1981). Taking all these arguments and counterarguments, Attfield concluded that the question of plants/trees having intrinsic value should remain an open one.

2.6.2.2 From "Reverence for Life" to "Respect for Nature"

From the "reverence for life" of Albert Schweitzer, we enter into a more inclusive and wider domain of "respect for nature" of Paul W. Taylor (Taylor 1981). Taylor advocated for a "life-centered system of environmental ethics" which differ from all anthropocentric ethical systems by virtue of its recognition of moral duties of humans towards plants, animals, and other members of the biotic community. Humans also have responsibilities for protecting these nonhumans and promoting their interests not merely for human good but for their own good. Therefore, not only humans are to be treated as "ends in themselves" (Johnson and Cureton 2022), all living things have to be brought under the ambit of this philosophy. Acceptance of such an ethical framework would lead to a change in the dominant moral attitude towards nature and strike a balance between the responsibilities of humans towards the well-being of their own species and towards that of the other members of the biosphere. However, in order to bring about this change in moral attitude, Taylor emphasized on the need to accept the concepts of "good" or "well-being, welfare" of a living being, and that of "inherent worth" (Taylor 1981). He further elaborated upon and explained these concepts in his book *Respect for Nature: A Theory of Environmental Ethics* (Taylor 1986). There, he stated that every individual organism, or populations comprising individual species, or an assemblage of different species to form a community, have its own good. This, in the case of non-human individuals or populations, comprises their full and healthy development, growth, reproductive success, abundance, density, etc. In the case of a community comprising many species, the maintenance of diversity and various community interactions could signify the good of a community. Humans must recognize that every nonhuman individual/population/ community – irrespective of whether it exhibits sentience or the ability to feel pleasure and pain – has its own good. The second factor that Taylor regarded as a prerequisite for developing a moral attitude that harbors a "respect for nature" is the concept of "inherent worth" (Taylor 1981), which humans would have to recognize in all nonhuman individuals, populations, and communities. He further clarified that his concept of "inherent worth" is comparable to that of "inherent value" of Tom Regan. Taylor also listed two "general principles" – "moral consideration" and "intrinsic value" – for identifying the inherent worth of an entity. The principle of moral consideration entails that all members of the Earth's biotic community deserve to receive compassionate and considerate treatment from humans (Taylor 1981). As said earlier, this consideration does not depend on the species to which the creature belongs, or whether it has sentience or not. According to the principle of intrinsic value, all members of the Earth's biotic community are "intrinsically valuable", or, in other words, deserve to be treated as ends in themselves, and not as possessors of merely an instrumental value by dint of which its value is only derived from its usefulness to another entity (primarily humans). However, an entity having an intrinsic value can also have instrumental value/s, and these are not in conflict with one another. Humans recognize inherent worth in other humans and consider them of value not just because of their social standing, or wealth, or merit, or other capabilities. The mutual respect between humans stems from this realization of the inherent rights and value of another human. It is, nevertheless, true that human history is replete with violations of this principle at different times and places, when slaves, people belonging to certain castes, communities, and color or gender, were evaluated mostly for their instrumental value. A strong or more skilled slave was more valuable than a less skilled or old and infirm one. In the same way, the dominant conventional ethic towards nonhumans only considers their instrumental value arising out of their usefulness to humans. Any nonhuman not instrumentally useful to humans is not considered to have any value. For example, plants growing on their own in flower gardens or crop fields are considered as "weeds", and removed manually or by applying poisonous chemicals. Unwanted animals are branded as pests and eradicated. To cite an example, "modern and scientific" forestry practices were initiated in India with the advent of British rule. This process began with the formulation of the first Indian Forest Policy in 1894. Dietrich Brandis, who is considered to be the founder of modern forestry in India, was well aware of the ecological values of forests. Despite this knowledge, he practiced the principle of "minimum diversity" and totally ignored the economically unimportant trees, shrubs, epiphytes, and creepers that make up the rich diversity of a tropical forest, and often play crucial roles in its ecological functioning. Brandis, on the other hand, put in place a policy of replacing them with economically valuable tree species for their active propagation (Rajan 1998). Thus, forest trees and other plants were only considered valuable for their instrumental value, and not any inherent worth. Hence, as emphasized by Taylor (1981, 1986), developing an environmental ethic that engenders a "respect for nature" is essential for a more ethical and compassionate treatment of nonhumans, which will have profound implications in resolving issues like biodiversity loss, deforestation, and climate change that impact humans and nonhumans alike.

Another relevant issue is that the "nonbasic interests" of humans are in conflict with his concept of "respect for nature" (Taylor 1986, 273), because pursuing these ends limits the value of nonhuman organisms to a mere instrumental type and denies the existence of any inherent worth. Examples of such nonbasic interests include (but are not limited to) killing of wild animals like elephants, rhinoceros, tigers, different fur-bearing animals, reptiles such as crocodiles, etc. for their body parts such as ivory, horn, nail, fur, skin, and so on. Trapping wild animals and birds to keep them as pets, hunting, and fishing for sport and recreation are also included among such activities. Humans who participate in such trade or sport as well as those who buy and consume such products do not have any "genuine respect for nature" (Taylor 1986, 274). The ultimate goal of ethicists and environmentalists would be to work for universal acceptance of such an ethic.

2.6.2.3 Peter Singer and Tom Regan – Sentience- and Right-Based Ethics

Peter Singer argued that the basic rationale behind the concept of equality was to take into account the "interests" of that particular being – "black or white, masculine or feminine, human or nonhuman" (Singer 1975, 34). Singer also made it clear that equality does not imply "equal or identical *treatment*". It means giving "equal consideration" (Singer 1975, 30). In his book *Animal Liberation*, Singer gave the reasoning that if differences in the degree of intellect do not permit some humans to exploit or suppress humans belonging to another race or sex, then extension of the same logic demands that humans do not enjoy the right to exploit nonhumans. He quoted Bentham who is among the few who said that one day the animals may be freed from human tyranny just as people with a different skin color or sex are being released from their subservience to their "superior" masters. The main criterion that Bentham cited is the ability to suffer. Therefore, Singer fixed the criterion of sentience as the benchmark for the ability to suffer, and hence, all sentient beings – human and nonhuman alike – should be given equal consideration. Sentience has been defined as "a multidimensional subjective phenomenon that refers to the depth of awareness an individual possesses about himself or herself and others" (Marino 2010). However, Singer also said that the "limit of sentience" and not "intelligence or rationality" is "a convenient if not strictly accurate shorthand for the capacity to suffer and/or experience enjoyment" (Singer 1975, 38). He elaborated on the question of sentience and of pain and cited our own observations, findings of animal welfare committees, and of neural and behavioral research to assert that "animals can feel pain" and the pain that they feel cannot be said to be less severe than that felt by humans. However, the threshold of pain may vary from creature to creature. For example, the force of a slap dealt to a child may not elicit a feeling of pain from a horse, but beating with a stick would cause it some pain. We may say here that this pain threshold also varies among individual humans, though all humans can be said to feel pain. In this way, Singer argued for recognizing sentience as the defining criterion for deserving moral consideration. If a given organism is not capable of feeling pain or enjoyment, then "there is nothing to be taken into account" (Singer 1975, 38). Singer also stated that inanimate objects like rocks, and plants including trees, could not deserve equal moral consideration with birds and mammals, because they lacked the capacity for sentience.

While Singer approached the question of granting moral consideration to animals on the basis of sentience, Tom Regan (Regan 1983) raised the issue of rights. He held human "conceit" and "chauvinism" to be responsible for preventing humans from recognizing the presence of consciousness in animals. Regan likened this chauvinism to the one that prevents the members of a particular race, religion, color, or gender from recognizing that the members of another group could also possess the same qualities possessed by them that they think are praiseworthy. Regan also deduced that all animals – at least the mammals – possess what he terms as "preference autonomy" (Regan 1983, 85). This he differentiates from the Kantian (German philosopher Immanuel Kant) definition of autonomy which may be defined as the ability to act rationally in "conformity with the laws that properly determine the will". Any action that violates these moral laws would be considered as "wrong". It is also obvious from this definition of autonomy that only a rational being can "act *according to his conception of laws*" (Brassington 2012, 167). Thus, in the Kantian sense, autonomous individuals can rise above their individual preferences and act in conformity with their moral duty. Regan reasoned that it is highly improbable that any non-human animal would be able to think at such a high level of reasoning. On the other hand, they do have preferences and are able to act to satisfy these preferences, for example choose the type of food or shelter that they prefer. Entities like rock, soil water, air, or even plants do not have either Kantian or preference type autonomy, but most animals – at least mammals – have the cognitive ability to have and satisfy "desires and goals". In the opinion of Kant, one must be autonomous in order to be a moral agent (Regan 1983,

85–86), and be accountable for doing something that is right or wrong. Nature cannot be a moral agent in this sense. However, there is also the category of "moral patient", which includes nonhuman animals, as well as "paradigm moral patients" (Regan 1983, 153) such as children and mentally ill or senile adults. These moral patients cannot become moral agents and therefore, cannot be morally judged for their actions, but deserve to receive moral treatment from moral agents. For example, Regan explained that beating a child or ill-treating a senile person is wrong, though not being moral agents, neither the child nor the senile person is accountable for their actions. Thus, moral agents have a mutual relation of doing right or wrong to each other, while in moral agent-moral patient relationships, moral agents are liable to treat moral patients in an ethical manner. Humans have moral obligations towards both paradigm moral patients (children and senile/ disabled humans) and animals. Regan also clarified that his was not a "speciesist" position, because moral patients like animals are excluded from the moral community not because they do not belong to a particular (human) species, but because they lack the cognitive abilities that justify their inclusion as moral agents. Both moral agents and moral patients (including animal moral patients) possess equal inherent value, which means "value in themselves". Being a "categorical value", it does not have grades, and is an all-or-none kind of property; either an entity has it or does not have. This implies that "all animals are equal" and upholds Albert Schweitzer's concept of "reverence for life" (Regan 1983, 235, 239, 241). However, Regan did not concur with Schweitzer's belief in not harming all that is alive, and argued that if simply "being alive" is sufficient reason for accepting direct duties, then "It is not clear why we have ... direct duties to, say, individual blades of grass, potatoes, or cancer cells". Instead, Regan introduced the "*subject-of-a-life criterion*" (Regan 1983, 243) for receiving moral consideration and justice, which is more than just being alive or just having consciousness. In order to be a subject-of-a-life, an individual must have the ability to feel "pleasure and pain", and have desires, perceptions, memories, and preferences. Only those entities satisfying these conditions have inherent value. Regan also introduced the "respect principle", which justifies the moral right of both moral agents and moral patients to "respectful treatment" (Regan 1983, 233, 279), since they possess equal inherent value.

Both Singer and Regan gave forceful arguments for extending environmental ethics beyond its anthropocentric confines to include animals – sentient by Singer, and all animals by Regan – though they fell short of including plants as well as ecosystems or landscapes within the domain of moral jurisdiction. Because of this inadequacy, their environmental ethic is ill-equipped to address complex, global issues involving both biotic and abiotic components of the environment such as climate change, acid rain, ozone layer depletion, and the like.

2.7 Tracing the Routes of Ecocentric Ethics

J. Baird Callicott attributed the development of ethics in human societies to two distinct lines of origin. One is the social contract theory of English philosopher Thomas Hobbes (1588–1679), which considers human societies as distinctly separate from animal societies, because the latter do not have true societies. In the early stage, humans existed as solitary individuals, who were in a state of constant strife with each other. Such an existence could not facilitate the development of collective enterprises such as agriculture or industry. This led to the development of social contracts, whereby moral codes and laws were formulated to make life better. Such a social contract emanated purely from a sense of "selfish rationality, [and] not selfless sentimentality" (Callicott 1999, 119). We could perhaps deduce from this that anthropocentric ethics, which only takes into account the interests of human societies, and in its worst form, does not even extend moral considerations to certain groups of humans such as slaves, people of color, certain gender or caste, could have links to such moral/ethical models. The second line of origin comprises "the sentiment-based moral philosophies" of Scottish philosopher David Hume (1711–1776), and Scottish economist-philosopher Adam Smith (1723–1790). Hume believed that humans are "born in a family-society", and the families frame and adopt the requisite rules for sustaining themselves. As several families come together to form a larger society, these rules get expanded, and in this way, "the boundaries of justice ... grow larger" as more societies are incorporated into a larger system (Hume 1777, 12–13). Darwin acknowledged this process, and wrote

> As man advances in civilisation, and small tribes are united into larger communities, the simplest reason would tell each individual that he ought to extend his social instincts and sympathies to all the members of the same nation, though personally unknown to him.

Darwin also reasoned that once such a state is reached, it is only a matter of time before human "sympathies" are extended to members of "all nations and races". He also held the opinion that "Sympathy beyond the confines of man, that is humanity to the lower animals, seems to be one of the latest acquisitions" (Darwin 1871, 100–101). This he surmised from the "abhorrent gladiatorial exhibitions" of the Romans.

Adam Smith also contended that however selfish humans may be, they still feel "pity or compassion" for "the misery of others". This compassion arises because humans imagine themselves to be in the place of the suffering person, and can feel the misery of that individual. Smith (1790, 4, 5) said that "... this is the source of our fellow-feeling for the misery of others" .... It may not be wrong, therefore, to deduce that an ethic founded on the principles of such "selfless sentimentality" culminates in ecocentric ethical principles, which is in constant opposition to anthropocentric attitudes and policies.

Aldo Leopold blended the essence of an "ecological world-view" (Callicott 1999, 121) into these models of ethical behavior, and extended the moral norms to take within their fold the entire land and its community, including humans living therein to give the broad dimensions of a land ethic, which is one of the first concepts of ecocentric ethics.

2.8 The Present Scenario and Possibilities for Ecocentrism

After about three or four decades of these debates, where do we stand today? As said earlier in Section 2.4, most international reports, conferences, conventions, and other documents still have an anthropocentric focus. Because of their limited success, the present situation demands that our environmental ethics need to be more inclusive and set to embrace both biotic and abiotic elements of the environment. Ecocentrism, being broadly inclusive, fulfills this condition, and could perhaps yield more positive results. Washington et al. (2017) have suggested the adoption of the broadest possible view of ecocentrism which includes in its fold biocentric (recognizes intrinsic value in all life forms), zoocentric (intrinsic value in animals), and geocentric (intrinsic value in geodiversity because it supports and sustains life) ethics. Such an ethic, therefore, presents the most inclusive concern for the environment. Many indigenous societies continue to follow a way of life that has ecocentric underpinnings, and a large section of environmental ethicists, conservationists, and other thinkers and philosophers have expressed their faith in an ecocentric worldview, making it feasible for its international recognition and acceptance. Washington et al. (2017) also refuted the allegation that ecocentrism is "anti-human". Ecocentrists simply see the humans as a part of the "ecosphere" and not as an entity existing outside its laws. Hence, humans do not have the liberty to manipulate it in an unchecked manner. Brian Baxter also contended that ecocentrism, while upholding the principle of justice for all humans, asserts that other species also have an equal claim to justice and a stake in the Earth's habitats and resources. He argued that nonhuman organisms have the right to live in the habitats required to enable them to thrive and reproduce in order to survive as a species. Institutional arrangements are needed to accommodate these claims of nonhumans and integrate them into human decision-making, citing the example of "German constitutional amendment to secure rights for animals". Ecocentrism can be said to provide a platform for "an acceptance of the moral considerability of nature". Thus, adoption of ecocentric ideals could go a long way in ensuring that no nonhuman species or population is denied access to vital resources – "without good moral cause shown" (Baxter 2005, 101–102, 137) – that are necessary for their continued survival.

2.9 Conclusions

The debates and discussions summarized in the preceding sections of this chapter suggest that the development of the fundamental ethical behavior had its origins in the confines of the family, and was gradually extended to include the clan, the tribe, the nation, and even the global human community. The growth of biocentric/ecocentric ethics took it beyond the confines of human societies to include sentient animals, then all animals, plants, and all forms of life, and even ecosystems where all the living communities carry out their life processes. Such an ethic also regards human communities as an integral part of this larger community. This is the essence of Leopold's land ethic, which was further developed to include myriad dimensions by many ecocentric thinkers. This ethic also acknowledges and draws its support and inspiration from the traditional "folk philosophies" and practices of indigenous societies, and the creative outputs of poets, writers, artists, and others. Inclusivity is its strength.

3

ECOCENTRISM AND THE QUESTION OF SENTIENCE IN ANIMALS AND PLANTS

3.1 Introduction

Peter Singer regarded sentience as the defining condition for granting of moral consideration to nonhumans, while Tom Regan introduced the concept of subject-of-a-life, where he argued that moral patients, which include all animals and some humans, have the moral right to receive respectful treatment from moral agents, which include all humans barring some special categories (see Chapter 2). However, even if the condition of sentience for granting moral consideration to nonhumans is accepted, we find that the definition and limits of sentience are changing with the rapid advancement of knowledge in neurobiology and cognition.

3.2 Sentience and Consciousness in Nonhuman Animals

3.2.1 Animals as Automata

René Descartes (1596–1650), the noted French mathematician and philosopher, believed that animals cannot reason and have no thought process. Though they may do certain things even better than humans, they do these like a clock, acting merely on their instinct, and not thinking. The lack of any thought process in animals is revealed by their lack of speech. He found that only humans possessed the two "principles" governing activity: one a "corporeal soul" and the other "the incorporeal mind", which is the soul that comprises "thinking substance". Animals only possess corporeal soul, but no mind. They are "automata" (Descartes 1976, 65–66) produced by nature. Though the Cartesian school of thought became a dominant one in western

DOI: 10.4324/9781003481058-3

society, there were people who believed otherwise. Voltaire (1694–1778) – French writer, historian, and philosopher – did not agree with Descartes and gave several examples of the ability of animals to learn from their experience and do things that reflect thinking on their part. He cited the example of a bird that could adjust the shape of its nest to fit to the surface to which it is attached; the dog that wails and is sorrowful when it loses its master, and is filled with joy when it finds him again. He reminded us that feeling can not only be perceived from speech, but from expressions and actions as well. Voltaire challenged the concept held by many philosophers that "Animals' souls are material", and asked that if it is a matter with sensation, then there is a need to know to what they owe this sensation (Voltaire 1976, 68).

3.2.2 Animals Have More Cognitive Abilities than Believed Earlier

More than three centuries after Descartes, Marian Dawkins questioned the practice of determining consciousness on the basis of higher cognitive abilities such as the ability to think of abstract concepts, making plans for future, and using language, which are used to determine the threshold of consciousness. Dawkins reasoned that such cognitive abilities are not required to experience basic emotions such as simple pleasure or pain. If the markers of consciousness are extended to include emotional attributes, then more animals could come within its ambit, rather than confining it to only those animals that possess higher cognitive abilities such as the ones mentioned earlier. It is possible that the "conscious experience of emotions" evolved much earlier than the higher abilities like concept-forming and using language. Dawkins argued that emotion deserves more attention in the study of consciousness. She referred to Bentham, who considered the capacity to suffer as the primary criterion and not the ability to reason and speak. Therefore, emotional awareness, which evolved long ago, could be regarded as the threshold of consciousness. However, she cautioned that various anticipatory phenomena, such as the ability of Dodder plants to select host branches having better nutrition, or choices made by chickens for a particular type of flooring, should not be taken as indicating possession of consciousness, because these could have evolved by natural selection and pre-programmed into their genes. As opposed to these responses, Dawkins cited the presence of reinforcement learning in nonhuman animals, which is the ability to change behavior through experience. For example, rats can learn that moving right leads to getting food, while moving left could make them receive an electric shock, making them to avoid moving left. When the situation is reversed, they can adjust their behavior accordingly by unlearning the previous experience and picking up the new one, that is, move left to get food and not go right to avoid electric shock. This is obviously subjective and cannot be an innate response. It is a "conscious awareness of pleasure and pain". Dawkins raised the question here that whether this simple experience of pleasure and pain could be regarded as the first conscious experience in the evolution of life, or whether consciousness is only characterized by the higher cognitive abilities. She drew attention to this question and left it open for resolving it further (Dawkins 2000).

In a study that primarily focused on the cognitive ability of domestic chicken (*Gallus gallus domesticus*), but also included an extensive review of scientific literature since 1990s on animal emotions, Lori Marino found that quite a few birds, mammals, fish, and invertebrates possess several cognitive abilities such as recognizing partly occluded (hidden) objects (monkeys, cats, several birds such as parrots, parakeets, macaws, magpies, crow, mynahs, jays, barn owls, chicken, and fish such as redtail splitfin and goldfish, and invertebrates such as bees); completely occluded objects (adult hens); numerical abilities (chimpanzees, orangutans, rhesus macaques, bottlenose dolphins, lions, elephants, horses, pigeons, crows, African gray parrots, and chickens); perception of time duration and anticipation of future events (chimpanzees and other great apes, bottlenose dolphins), simple time perception (pigeons, black-capped chickadees, and scrub jays); episodic memory of remembering past experiences (many mammals, including great apes, birds such as pigeons, western scrub jays, chicks and adult hens); self-control (mammals such as rats, lemurs, rhesus macaques, chimpanzees and orangutans, birds like pigeons, black-capped chickadees, carrion crow, common raven, and chickens); applying logic and reasoning (chimpanzees, various species of monkeys, rats, several species of birds); self-assessment (birds such as graylag geese and pinyon jays); using cognitive skills that indicate cognitive complexity and intelligence (domestic pigs, dogs, primates, dolphins, whales, several birds); individual discrimination in social groups (dogs, pigs, elephants, vervet monkeys, macaques, dolphins); vocal recognition (various species of songbirds); visual recognition of conspecifics (rooks, pigeons, white-throated sparrows, chickens, and budgerigars); odor recognition among conspecifics (Antarctic prions); perspective taking and social manipulation (chimpanzees, dogs, pigs, western scrub jays); emotion-based cognitive bias (rats, pigs, primates, birds like starlings); and personality traits (several species of fish, birds such as zebra finches, great tits, graylag geese, Japanese quail, and mammals). It is apparent that the list is fairly long, and the abilities are diverse (Marino 2017).

3.2.3 Neo-Cartesianism versus Evolutionary Continuity

The school of Neo-Cartesianism (NC) and that of Evolutionary Continuity (EC) hold opposing views on animal sentience and consciousness. Several Neo-Cartesians (NCs) including both religious and non-religious thinkers have been forwarding their views since the 1960s. They advance the idea that pains felt by animals do not qualify for moral consideration, the major defense for which comes from the differences between animal and human brains. Human brain has an enlarged prefrontal cortex (PFC), which is not so remarkably developed in many animals. For example, fish lack the anterior cingulate cortex (ACC), and therefore, do not have pain perception (Halper et al. 2021). Brian Key put forward arguments to support the lack of pain perception in fish. To state in a simple way without delving deep into the neurobiological details, pain originates in humans through neural activity in the cerebral cortex on the anterior brain surface. Several cerebral cortical regions such as the prefrontal cortex, anterior cingulate cortex, somatosensory areas I and II, and the insular cortex become active during pain in humans as shown by functional magnetic resonance imaging (fmri). An important neuroanatomical feature contributing toward the ability to feel pain is the parcellation of neural tissue into regions that can execute various computations related to pain. Fish lack such parcellation and many other features including essential networking necessary for feeling pain (Key 2016). Rose et al. (2012) also came to the conclusion that on the basis of behavioral and neurobiological evidence, fishes have very limited nociceptive responses, and are unlikely to experience pain. The C fiber nociceptors – which are responsible for feeling of pain in humans – are absent in elasmobranchs while these are rare in teleosts. Contrary to this observation, Sneddon et al. (2014) in their review of available literature cited several works to show that the forebrain and midbrain regions in teleosts are active when they receive painful stimulus, though they do not behave similarly during innocuous treatments or exposure. These responses include gene expression, and/or electrical activity, and/or fMRI (functional magnetic resonance imaging)-detected metabolic activity, in a wide range of fishes such as carp, rainbow trout, different salmon species, and goldfish. Several studies also provide support for the contention that anterior cingulate cortex (ACC) and insular cortex are not strictly essential for experiencing pain effects (Halper et al. 2021). However, it may be pertinent to mention here that many animals including humans have a nonconscious processing of stimuli to prevent tissue damage. This perception is known as nociception. One common example is withdrawing our hand when we accidentally touch a hot surface. The feeling of pain appears later. When humans talk about their experiences of pain, they essentially talk about the conscious phase of pain perception, which comes after nociception. Therefore, it would be wrong to confuse nociception with actual pain perception. Nociceptive or "nociceptive-like" abilities go pretty far down the evolutionary tree, at least up to the Cnidarians, which include animals like jellyfish, corals, and sea anemones (Braithwaite 2010, 36). In her book *Do Fish Feel Pain?*, Victoria Braithwaite presented some findings (Braithwaite 2010) which are contrary to what Rose et al. (2012) had found. She reported the results of some experiments conducted by her and her co-researchers on the ability of fish to feel pain. They examined the trigeminal nerve which in vertebrates supplies the maxillary (upper jaw) and mandibular (lower jaw) areas. It transmits stimuli from an injury via the spinal cord to the brainstem via two types of nerve fibers: i) the A-delta fibers, which are larger, insulated with myelin sheath, transmit faster, and are meant for transmitting "first pain"; and ii) the C fibers, which are thinner, not insulated, transmit at a slower rate, and are responsible for transmitting "second pain". Braithwaite and her co-researchers found the presence of both A- and C-type fibers in the trigeminal nerve of trout. However, the proportion of C fibers was ~4%, which is much lower than the 50-60% found in other vertebrates. They also found the presence of nociceptors in the mouth and facial area of the fish, which when subjected to noxious stimuli such as touch/heat/chemical (vinegar) triggered an electrical signal that was transmitted via the trigeminal nerve into the brain to alter the behavioral patterns of the fish. Trout injected with bee venom and/or vinegar in their facial areas under the skin show higher respiration rates and reduced hunger, which are believed to be the initial evidence for the presence of pain receptors. However, this does not constitute conclusive evidence, because these can simply be primary physiological responses and not resulting from a cognitive experience. To give more conclusive evidence, Sneddon et al. (2003) conducted the "novel object test", which comprises introducing a new object (in this case, a brightly colored Lego brick tower) in the experimental tank. Trout is known to avoid any such new object till it can find out that it is not a threat. This is known as neophobia. In the experiment, trout injected with either saline (control) or 2% acetic acid were introduced into the tank containing the novel object, which elicited different responses. The saline-injected control trout showed the normal avoidance reaction, along with an increase in respiration rate. In contrast, the acid-injected fish moved very close to the object and did not have increased respiration rate. This may be interpreted as impairment of attention due to the "pain" caused by vinegar exposure in trout. In a subsequent experiment, when both saline-treated and acid-treated fish were given a dose of morphine which acts as a pain reliever, the acid-treated fish also started showing the normal avoidance behavior. This showed that alleviation of pain by morphine enabled the fish to regain its normal behavior. Since avoidance is a complex behavior, this also indicates that fish could cognitively recognize tissue damage, or in other words, it feels "pain". Brian Key – a strong advocate of the view that fish cannot feel pain – has also suggested that "pain-suppressed rather than pain-elicited" normal behaviors such as feeding, locomotion, etc. need to be studied more to understand the perception of pain, instead of focusing on "reflex behavior" such as "escape response" (Key 2016). In their commentary on Key (2016), Braithwaite and Droege (2016) pointed out that adopting human pain processing as the model for pain perception in all animals is not the right approach, because it is possible that convergent evolution in diverse groups of animals could have resulted in different structures performing similar functions. Donald M. Broom's commentary on Key (2016) also suggests that fish brain and behavior indicate their ability to feel pain (Broom 2016). Broom is of the opinion that nociception and pain should not be viewed as two distinct responses, but as a part of the entire "pain system". He also reviewed the earlier works and concluded that different species/groups of fish might use different parts of the brain to give responses to fear and pain. However, these parts have functions that are analogous to those of mammalian amygdala and hippocampus, which are brain areas that process emotion and feeling. In fish, the areas responsible for pain and fear responses may include habenula (complex nucleus connecting limbic forebrain and midbrain) and optic tectum.

3.2.4 Intelligence in Fish

Culum Brown in his extensive review of published literature during 1960–2014 on the intelligence and cognitive abilities of fish found that fish are far more intelligent than they are commonly thought to be. They are capable of several sophisticated behaviors and have long-term memories and individual recognition (Brown 2015). They also show signs of "Machiavellian intelligence" (Bshary 2011) and other advanced behavioral elements. Named after the 15th–16th-century Italian politician and writer Niccolo Machiavelli, Machiavellian intelligence involves manipulation of group members, deceitful behavior, and cunning strategies to gain advantages in social groups. It is known to occur in higher primates and humans and can be used to exercise power in social groups (Byrne and Whiten 1997; Knecht 2004). Development of Machiavellian intelligence requires that an individual knows all other group members and their relationships with other members and utilizes this information to not only build successful coalitions but also prevent opponents from building coalitions. Besides humans and primates, many bird species are now known to possess such intelligence. Fishes also possess a number of cognitive abilities that enable them to elicit information by observing the interactions between other individuals in the group. This also makes it possible for them to adjust and adapt their behavior to derive maximum advantages within the group (Bshary 2011). Therefore, it may not be wrong to say that if the various advanced faculties of the fish brain are taken into account, the dissimilarities between human and fish brain turn out to be less than what would superficially appear, and fish are likely to be conscious and capable of feeling pain. Many fish species have tetrachromatic vision, that is, they use four cone photoreceptor classes for color vision (humans use three), and have good visual acuity and olfactory ability. Cerebral lateralization (division of labor between the two cerebral hemispheres: for example, the left is used for language production, while the right for assessing visual and spatial relationships between objects), which was earlier thought to be unique to humans, is also present in fishes. For example, Sarasins minnows (*Oryzias sarasinorum*) uses its left eye to look at familiar individuals, while it uses its right eye to look at unfamiliar individuals, including predators. Various types of cerebral lateralization are also seen in other fish species such as rainbowfish (Family: Melanotaeniidae). Based on these evidences, Brown (2015) contended that the sensory abilities of fish are also highly developed, and comparable to humans in terms of several abilities. Hence, it would be unfair to say that fishes do not have cognitive abilities and are dumb creatures. Fish have remarkable memory and are known to remember the persons who feed them regularly, even when the person comes to feed them after a break of six months. Fish also remember the time and the probable location where they are usually fed based on their earlier experience. Thus, they possess an ability called "time-place" learning, which is the ability to associate the time of feeding and the place where food is likely to be given. The time taken to learn to associate time and place ranges from 2 to 4 weeks in different fish species/families, which is comparable to the time of 19 days taken by rats to learn the same. When goldfish and rainbow trout were provided stimuli with brush and pin-prod to elicit responses in spinal cord, cerebellum, tectum, and telencephalon, the nociceptive stimuli with pin-prod elicited greater neuronal activity than mechanoreceptive stimuli with brush in goldfish, but this was not the case in trout. The study also revealed that both A-delta and C-type fibers are activated (Dunlop and Laming 2005). These findings suggest that there is a nociceptive pathway to higher brain centers in the central nervous system of fish, and therefore, a potential for pain perception. Further, goldfish and trout have the ability to learn to avoid noxious nociceptive stimuli such as both low- and high-voltage electric shock, and this is not an innate reflex action because they can change the innate response to adapt to changing situations. These findings show that fish could have learned pain perception (Dunlop et al. 2006). Braithwaite and Ebbeson (2014) found that fish species which are farmed can also feel stress and pain, and different species show varying abilities to avoid stress. Farmed fish also experience pain relief from the effects of acetic acid injection after the application of carprofen, a non-steroidal anti-inflammatory drug, and a local anesthetic lidocaine. These findings throw light on pain perception in fish and have important implications in on-farm fish welfare. Yew-Kwang Ng in the commentary on Brian Key's paper commented that Key (2016) failed to provide conclusive evidence for proving that fishes are not able to feel pain (Ng 2016). In fact, several scientists believe that the capacity to feel pain evolved fairly early, and is present in all mammals as well as all vertebrates. Many studies show that the neural basis for this capacity lies in the lower part of the brain and not exclusively in the cerebral cortex. Therefore, animals like fishes that do not have a cerebral cortex are not necessarily deprived of pain perception. It is, therefore, now necessary to take a wider perspective of pain and its perception. Sneddon and Leach (2016) responded to the skeptics who insist that fish or any animal other than primates cannot feel pain because they lack a laminar cortex and that fish does not have C fibers which convey noxious stimuli to the brain though it has been found that A-delta fibers can perform the function of C fibers. Moreover, C fibers have been shown to be present in certain fishes such as the trout, albeit in lesser proportions when compared to other vertebrates (Braithwaite 2010). Furthermore, fish show long-term responses to pain, restore normal behavior after application of analgesics, and other such pain-related responses. Brown (2016) also shared this view and said that the attempt to define pain on the basis of a structure-function relation by solely attributing the capacity to feel pain to the presence of the cerebral cortex ignored the evolutionary trend of gradually moving from one stage or level to another and not in "all-or-nothing leaps". The cortex in humans has taken up many functions which were earlier performed by other regions in the brain. The view that it is absolutely necessary to have a cortex in order to have pain perception also ignores the role of convergent evolution whereby different structures in different groups can play a similar or the same role. Moreover, nociception and emotional response to pain could be regarded as parts of the same pain system (Broom 2016). These, and many other findings, recommend moving away from a narrow, human-centered concept of pain perception in the interest of not only factual accuracy but also animal welfare.

3.2.5 Relevance of Precautionary Principle in Animal Sentience

Jonathan Birch invoked the precautionary principle in the debate over animal sentience (Birch 2017). The precautionary principle is contained in Principle 15 of the "Report of the United Nations Conference on Environment and Development" (1992). The principle says that

> In order to protect the environment, the precautionary approach shall be widely applied by States according to their capabilities. Where there are threats of serious or irreversible damage, lack of full scientific certainty shall not be used as a reason for postponing cost-effective measures to prevent environmental degradation.

This report was subsequently adopted in the UN General Assembly on August 12, 1992 (United Nations 1992a). Speaking in a historical perspective, the basic tenets of the precautionary approach have existed since long. It lies at the root of ancient public health principles such as "first do no harm" and "an ounce of prevention is worth a pound of cure" (Raffensperger and Tickner 1999, 1). However, it was perhaps first adopted officially in the "Ministerial Declaration Calling for Reduction of Pollution (November 25, 1987)" of the Second North Sea Declaration (Raffensperger and Tickner 1999, 356). The Wingspread Declaration of 1998 also reiterates that "When an activity raises threats of harm to human health or the environment, precautionary measures should be taken even if some cause-and-effect relationships are not fully established scientifically" (Raffensperger and Tickner 1999, 8). The precautionary principle has found its relevance, potential, and actual application in several areas such as toxics pollution of the environment, ozone layer depletion, environmental and human health, sustainable agriculture, and approval of new and potentially hazardous technologies. It has become a major guiding principle of the European Union (Kriebel et al. 2001). In the light of the aforementioned findings and reasonings, Birch (2017) found its application relevant in animal sentience because even if there is a lack of scientific certainty about the sentience status of an animal and whether it feels pain, the benefit of doubt should go to the animal. Animal protection legislation should, therefore, extend to include all vertebrates and some invertebrates such as cephalopods. This also underscores the importance of increased recognition of ecocentric principles and worldviews among environmental/animal welfare policy-makers in order to facilitate the framing of more nature-friendly and compassionate policies.

3.2.6 A Revised Concept of Pain

A recent review has called for a rethink on the human-centered concept of pain in nonhuman animals (Schroeder 2018). In this review, Paul Schroeder emphasized the need to expand this concept to include other responses such as motor actions, avoidance behavior, and the presence of awareness of some harm. He recounted the major points that indicate the presence of pain in animals (Bateson 1991, as cited in Schroeder 2018; Sneddon et al. 2014). If an animal possesses nociceptors sensitive to pain, has structures analogous to the human cerebral cortex, has neural connections between nociceptors and brain, has opioid receptors in its CNS (central nervous system, i.e., brain and spinal cord), experiences reduction of noxious effects after administering analgesics, has the ability to learn to avoid painful stimuli, and this learning is "rapid and inelastic", then it can be said to have the ability to feel pain. However, the way a fish perceives pain may be different from that in humans, and therefore, it is not justified to judge fish pain on the basis of human criteria. Physiological and behavioral responses such as increased swimming activity, mucus secretion on gills, increase in plasma cortisol, and escape response to heat, observed in different species of fish, may also indicate the presence of pain. These responses may, however, differ among different species or groups. For example, zebrafish and trout respond to nociceptive stimuli by increasing their ventilation rate, but the same is not true of carp. Again, swimming rate decreases in zebrafish, but increases in rainbow trout. Zebrafish injected with acetic acid near tailfin exhibit "tail-fin fanning". Thus, the view that fishes lack the neural structures to have consciousness or feel pain based on mammalian neuroanatomy may not be an appropriate way of judging the pain perception abilities of fishes. This is true for other non-mammalian groups as well. For example, the pallium in birds is now believed to serve cognitive functions similar to that being done by the neocortex in mammals.

3.2.7 Consciousness in Nonhuman Vertebrates

Fabbro et al. (2015) have given a broad overview of available information on the state of consciousness in different groups of vertebrates. They suggest that animals such as fishes, whose nervous system mainly comprises of the basal subcortical system, have developed a primary consciousness. It may be relevant in this context to note that the return of persons under general anesthesia to normalcy is mainly characterized by brainstem and diencephalic activity. They also state that the role of basal subcortical structures is significant in developing self- and world consciousness, as has also been shown by several studies. Subcortical neural circuits are adequately developed in fishes and in other vertebrates. Several other midbrain and brainstem structures such as the hypothalamus and the roof of the midbrain are involved in developing a primary consciousness in vertebrates. Based on the ability of many fish species to choose and discriminate between partners, recognizing own odor, possessing long-term memory, and assessing own status and making strategic decisions and "third-party assessments", they can be said to have a "primary cognitive consciousness", with some recognition of a "proto-self" and possibly a "core-self" and an assessment of the world around them (Fabbro et al. 2015). The concepts of proto-self, core-self–core consciousness, and autobiographical self-extended consciousness were originally proposed by Antonio Damasio (Damasio 1998, 1999). Core consciousness is the basic type of consciousness that Damasio (1998) also called "awareness", while the more advanced extended consciousness – which is dependent on core consciousness – could also be termed simply as "consciousness" (Damasio 1998, 1880). Core consciousness is the awareness of existence at a given moment and a given place. It is shared by many nonhuman species with humans, and is associated with the "transient" core-self that is "re-created" with every interaction of the brain with an object (Damasio 1999, 17). Extended consciousness is a more complex type of awareness that is associated with an autobiographical self that evolves over time with accumulation of memories and experiences of the past and anticipations for the future. Extended consciousness is thought to reach its zenith in the human species and is greatly facilitated by the development of language. Nevertheless, extended consciousness may also be present in some nonhumans, albeit at simpler levels. It must also be remembered that the two types of consciousness are connected; extended consciousness "is built on the foundation of core consciousness" (Damasio 1999, 17). However, neither the core-self nor the autobiographical self is the original type of self that evolved. Damasio (1999, 153–154) proposed that a "preconscious biological precedent" – the proto-self – is the "nonconscious precursor" to the core-self. However, Gonzalo Munévar criticized the concept of proto-self, core-self (consciousness), and extended consciousness of Damasio and said that they suffered from serious defects. He further argued that the concept of core consciousness cannot explain phenomena which do not involve handling of an object such as during dreaming or locked-in syndrome (Munévar 2014). The latter is a neurological disorder in which the brainstem is damaged, resulting in paralysis of voluntary muscles due to which the affected persons are not able to speak, move, or even show facial expressions except moving their eyes up and down but not from side to side. They are, nevertheless, conscious and have normal cognitive abilities such as thinking and reasoning, but cannot handle any object. Anyway, coming back to the issue of consciousness in nonhuman vertebrates, the optic tectum (the largest and most conspicuous visual center in non-mammalian vertebrates) in the frogs may be mainly responsible for the representation of the world and a proto-self. Frogs can use visual cues to distinguish between a prey and a potential predator. However, they do not show any stress-induced emotional responses like fever or tachycardia (rapid heart rate). Nevertheless, based on the fact that the optic tectum is highly developed in amphibians along with the presence of telencephalon elements, it is probable that they possess a primary consciousness as well. The brain volume about doubles in reptiles from amphibians, and they show aversion to food injected with a chemical (lithium hydrochloride) that induces nausea. This is also seen in birds and mammals, suggesting that they feel "pleasure and displeasure". These three groups also develop fever and tachycardia upon handling. Some reptiles learn to open a door for getting food after watching another individual of the same species doing the same. Some species also show signs of REM sleep (rapid eye movement sleep – less deep sleep when the brain is active and one can have dreams), suggesting the presence of some form of primary consciousness. Moving from reptiles, birds possess a highly developed telencephalon or forebrain. Equipped with a highly developed sense of sight, birds are able to construct "complex spatial maps of the world" with the help of the ability of their brains to have "multisensory integration". Their learning, emotional, and cognitive abilities are highly developed, and they exhibit REM as well as slow wave sleep. Thus, they are likely to be able to dream and have "phenomenal consciousness". Some birds such as the European magpie (*Pica pica*) have developed self-consciousness as shown by their ability of "mirror self-recognition". This they share with – besides humans – with a few mammals such as chimpanzees, orangutans, dolphins, and elephants. Thus, all vertebrates including fishes have an awareness of the "primary self", while consciousness about the "core self" is present in reptiles, birds, and mammals. There is an ongoing debate about whether insects and other invertebrates such as cephalopod mollusks and decapod crustaceans have consciousness. Many scientists accept the presence of insect consciousness, with many others in opposition (Birch 2022). In fact, Edelman and Seth (2009) suggested that both birds and cephalopod mollusks show complex behavior and have highly developed nervous systems. Birds have sophisticated food capture behavior, memory, tool use and manufacture, and vocal learning, which stand testimony to their neural capabilities. The octopus has a large concentration of sensory receptors. The brains – especially the optic lobes – of octopus and the squid *Loligo* show remarkable development, and the cephalopod brain may have structures equivalent to the vertebrate forebrain. These findings raise important questions about non-mammalian consciousness. Several animal groups such as fish, amphibians, reptiles, and some invertebrates, which lack large brains and complex behaviors, may experience forms of conscious pain (Walters 2016). Pain-capable neural substrates may be widely distributed in the animal kingdom, though the networks in these animals may be much smaller than the cortical pain matrix in mammals. Hence, this pain may be qualitatively different from the way humans feel pain. It may be time to extend our empathy net wider and come out from the anthropocentric confines of our thoughts. Many years ago, Russell Church in 1959 conducted experiments to show that rats were empathetic to pain in other rats. This encouraged more studies on "empathy, sympathy and altruism" in animals. However, such studies are discouraged by the prevalent behaviorist attitude (de Waal 2007/2008). Social interactions play an important role in the development of empathy in animals, as it is observed that mice show empathetic behavior only with their cage mates (Langford et al. 2006). Humans need to accept that just as evolution – both gene and culture – has bestowed in them the "human way", it has similarly given an animal (or a plant) its own "way". Animal welfare need not come down from a high pedestal but has to respect an animal's naturalness which embodies its intrinsic values. Many animals can become a part of the moral community and have moral importance in their own right, and not in human interests or – as said by some philosophers – in order to keep our own humanity intact (Musschenga 2002). Thus, the findings discussed in this and the preceding sections of this chapter on the likely presence of consciousness in nonhuman animals – perhaps in a form different from that in humans – vindicate the ecocentric principle of recognition of intrinsic value in animals. This would automatically bring them under the ambit of a "moral coverage net".

3.3 Consciousness and Sentience in Plants

3.3.1 Can Plants Feel Pain?

Picking up the threads from the yet unresolved questions of pain and consciousness in non-mammalian vertebrates and invertebrates, František Baluška showed that plants, though not regarded as behaviorally active, produced chemicals having pain-killing and anesthetic properties under stress. Furthermore, sensitive plants treated with anesthetics cannot fold their leaves when touched. Plants like Venus flytrap cannot close their traps to capture animal prey under the influence of anesthetics (Baluška 2016). Though not yet universally accepted, plants are now visualized as "sensory and communicative organisms", which also exhibit active behavior and problem-solving abilities. Plants are rooted in one place, but they are not sedentary; they show active movements of their body parts, albeit at a slower rate than that of animals. Studies conducted in plant molecular and cellular biology, electrophysiology, and ecology show that plants not only show sensory-based behavior, they also possess "prototypic intelligence" (Baluška et al. 2009).

3.3.2 The "Root-Brain" Hypothesis of Charles and Francis Darwin

It is interesting to note that Charles Darwin conducted botanical researches in the later part of his life. These studies provided keen insights into the life of plants. Darwin along with his son Francis Darwin published a landmark volume entitled *The Power of Movements in Plants* in 1880 (Darwin 1880/2009). In Chapter III of this book, Darwin described the results of several experiments on the sensitivity of the root apex of some plant species upon contact with a surface and several irritants. The Darwins also showed that the root cap has the ability to move away from the point of contact or opposite to the side on which the irritant such as a chemical or high temperature or abrasion is applied. The roots could even defy geotropism (the tendency of roots to move downward) for some distance to avoid the obstructing surface or the irritant stimuli. Different plant species differed in their sensitivity to contact and irritants. For example, bean and pea root apices were highly sensitive to contact; that of *Phaseolus* and cucurbits more sensitive to caustic and abrasion than to contact; and that of the maize highly sensitive to both contact and caustic. The sensitiveness was confined to the tip of the radicle for a length of 1–1.5 mm, and this was transmitted to the upper part inducing it to bend. They inferred from these and other observations that the tip of the root guides it to direct its course through the ground, and in so doing "acts like the brain of one of the lower animals;

the brain being seated within the anterior end of the body, receiving impressions from the sense-organs, and directing the several movements" (Darwin 1880/2009, 573). However, Darwin's observations and inferences were not accepted by prominent botanists such as Julius Sachs, who found fault with the experimental methods of the Darwins, though it was later shown that it was Sach's and not Darwins' experiments that suffered from methodological errors.

3.3.3 Experimental Evidence for "Root-Brain" Hypothesis

More than a hundred years later, Darwin's concepts of the "root-brain hypothesis" found support in the works of Baluška and his co-workers. They found the existence of a transition zone between the root cap zone and the elongation zone. This transition zone served as a "brain"-like command center, as hypothesized by Darwin. This finding also vindicated the Darwinian concept of a common descent of all organisms – both plants and animals – and the "unity of life" postulate of Sir Jagadish Chandra Bose (Baluška et al. 2009). As roots go deeper into the soil, their tips sense the various objects in their route and respond either positively or negatively to numerous physio-chemical and biotic variables present in the soil, depending on the nature of these variables. However, the tips of the roots are essentially the sensory areas, whereas the sites for motor action are further up in the root. This implies that plants also have a "sensory-motoric circuit" (Baluška et al. 2009) as in animals. In other words, the growing root tips essentially perform invertebrate animal-like exploratory crawling movements to find the best course. Plants have the ability to gather information about their abiotic environment – especially light and gravity – and can store information as well as bring it back from memory. The transport of the plant hormone auxin across the cellular framework makes possible the translation of these information into various cellular activities. Plants have specialized cells, for example, statocytes in root caps and cells in the root transition zone that are capable of translating sensory information collected by the root tip to motoric responses in the elongation zone. Based on the sensory information collected by the root tips, plants can move straight or bend in a given direction (Baluška and Mancuso 2009). Plants have also been shown to have swarm intelligence, which is a mechanism that solves cognitive problems through the processing of information gathered by two or more individuals through social interactions. Such intelligence arising out of group living has been shown to be operative in humans and many animals including social insects. However, swarm intelligence is not just living in groups, it is solving problems together, based on group inputs (Krause et al. 2009). Plants have complex and extensive root systems, and the growing root apices exchange different types of information through their "neuronal-like" networks, chemicals and volatiles, as well as electrical communication via the electrical field developed around root apices. Thus, there is evidence for the existence of swarm intelligence in plant roots, which enables them to chart their courses through the soil and to optimally exploit nutrients and other soil resources. The root apex region plays an important role in receiving various environmental stimuli. The root cap is a sensory organ that perceives information regarding variables such as gravity, light, oxygen, various nutrients, and so on. Furthermore, there is a transition zone that serves as a command center and is located between the root cap and the elongation zone. Cells of the transition zone receive sensory information from the root cap and transmit them to the elongation zone to control the motoric activities of the cells there. This contradicts the age-old view that plants are passive organisms without any sensory perception and control (Baluška and Mancuso 2010).

Despite these findings, the integrating principles of plant neurobiology have met staunch resistance from many quarters. Baluška et al. (2009) have referred to the observation that many tenets of plant neurobiology evoke complex, diffuse, and often poorly understood ideas which violate the "parsimony principle" or the Ockham's razor – which owes its name to the 14th-century Franciscan monk William of Ockham. The razor requires that a valid theory is based on the simplest of ideas and with the least number of assumptions (Struik et al. 2008). However, here Baluška et al. (2009) also invoke Ockham's broom, which is a relatively new concept introduced by the Nobel laureate molecular biologist Sydney Brenner. The Broom represents the practice of pushing inconvenient facts under the rug to justify a long-standing dogma (Robertson 2009), which in this case is the air-tight barrier between the plant and the animal world held since the time of Aristotle.

3.3.4 The History of Plant Neurobiology

If we take a look at the history of plant neurobiology and sensory perceptions of plants, a deep-seated dogma accompanied by subtle and not-so-subtle "racism" and "eugenicist" thinking among a group of influential scientists largely contributed toward the denouncement and discrediting of the findings of Sir Jagadish Chandra Bose, who was a pioneering plant neurobiology researcher from India, and who is now regarded as the "father of plant neurobiology" (Minorsky 2021).

Sir Jagadish Chandra Bose (1858–1937) was a well-known Indian scientist, who in the early part of his research career was an experimental physicist. He is credited with his work on producing microwaves in the laboratory and characterizing their properties. He is also known to be a pioneer in the fields of radio engineering and semiconductor research. However, in his mid-career, he shifted his attention to plant physiology, where he studied the movements of the leaves of the sensitive "touch-me-not" plant

Mimosa pudica and the telegraph plant *Codariocalyx motorius* (formerly *Desmodium gyrans*). He later extended his findings on these two plants to other, more common, plants which do not show the motile behavior characteristic of these two species. Bose observed that the folding movement of the leaves of *Mimosa pudica* when touched traveled downward with each pair of leaves closing in succession. This movement resembled the action potentials (APs) generated in an animal during the course of a nerve impulse. In the case of *Codariocalyx motorius*, the periodic up and down oscillatory movements of its lateral leaflets appeared to Bose to resemble that of a pulsating heart. Minorsky (2021) quoted from a letter written in 1926 by Daniel T. MacDougal, Director, Desert Laboratory, Tucson, Arizona, and one of the chief detractors of Bose in USA: "his [*Bose's*] pulsations, heart-beats and nerves in plants are sheer nonsense with no scientific foundation whatever". He further said that Bose's theory of the ascent of sap was "utterly lacking in scientific significance", and "The heartbeats of plants which Sir Jagadis Chunder Bose claims he has demonstrated are mere figments of romantic oriental imagination, unsupported by any genuine scientific fact". The discrediting of the findings of Bose by an influential scientist like MacDougal pushed Bose's findings to the backstage and was branded as "pseudoscience".

After nearly a hundred years of Bose's experiments, Minorsky (2021) while analyzing the scientific integrity and accuracy of Bose pointed out that being an experimental physicist, Bose could design several instruments for measuring the autographs of plants. Of these, the finest and the most efficient was the magnetic crescograph, which could record at a mind-boggling magnification of 1 to 10 million times. With this equipment, Bose showed the generation of APs in plants just like that in animals. More recent studies reveal that plant APs involve calcium ion-activated channels, while animal APs use voltage-dependent sodium ion channels, indicating that APs in plants and animals reflect the process of convergent evolution. The plant cells that propagate AP are not structurally identical to the animal neuron, but if their functional roles are considered, then plant cells that can propagate all-or-none electrical impulses can be said to constitute a "diffuse nervous system" (Minorsky 2021). Bose was fully aware of the fact that the pulsating "heart" that he found in the Indian telegraph plant was not really a multi-chambered animal heart, but a simple primitive organ as found in some invertebrates. What is unique is that Bose found the oscillations of the lateral leaflets of the telegraph plant to be accompanied by electric oscillations in their pulvini, that is, the enlarged part at the base of their petioles. When he took the pulvini and immersed them in water, the oscillatory movements ceased, but were restored by adding hydrogen peroxide. The oscillations were also found to be inhibited by factors such as anesthetic agents, low levels of oxygen, metabolic inhibitors, and low temperature. Most significantly, atropine and pilocarpine – known for their antagonistic and agonistic effects, respectively, on acetylcholine receptors in animals – produced opposite effects on animal heart. These chemicals also have opposite effects on the leaflet movements of *Codariocalyx motorius*. Bose found that the inner cortex produced similar oscillations in other plants as well. Later studies showed that electric potentials arise in the cortex of plant roots, and are propagated along the xylem/parenchyma interface (Toko et al. 1990). Though these feeble pulsations could not drive the large volume of water transported along the plant's axis as believed by Bose, it does not mean that they do not exist. Hence, the charge of MacDougal that these pulsations are figments of oriental imagination is not correct at all. Therefore, Bose can be said to have laid the foundations of present-day plant neurobiology (Minorsky 2021). Later research has also thrown light on the ability of plants to respond to anesthetics and ion blockers. De Luccia (2012) found that amlodipine (calcium channel blocker) and ketamine (NMDA receptor) did not have any effect on the sensitive plant *Mimosa pudica* and Venus flytrap *Dionaea muscipula*. NMDA (N-methyl-D-aspartate)-receptors belong to a family of L-glutamate receptors, and play a very important role in memory and learning. Lidocaine (sodium channel blocker), on the other hand, has a significant effect on the sensitive plant (*Mimosa*), but not on Venus flytrap. Diethyl ether, an anesthetic, has significant effects on both the plants. Both lidocaine and ether have toxic effects such as dehydration, eventually leading to death. Grémiaux et al. (2014) recalled the pioneering studies of the French scientist Claude Bernard (1813–1878) who found that when animals were anesthetized with volatile anesthetics such as ether, they first lost their consciousness, though their vital functions were not affected. With longer exposure, the nervous system was disabled and respiratory movements ceased, and in the next stage, heartbeats and ciliary movements were also disrupted. Bernard went on to show that in a similar fashion, plant cells also responded to volatile anesthetics, whereby movement ceased first, although circadian movements of leaf was not disrupted (as shown in *Mimosa pudica*). Seed germination and photosynthesis were also affected by ether exposure. Thus, Bernard showed that plant cells were responsive to anesthetics just like animal cells, and just as neurons in animals are more sensitive, plants also have cells which are more sensitive to anesthetics. This finding, therefore, challenged the long-held notion that animal life was based on "senses and movements", but not so in plants, and provided support for the theory of "unity of life".

3.3.5 The Question of Plant "Intelligence"

Sternberg (2022) has defined human intelligence as "mental quality that consists of the abilities to learn from experience, adapt to new situations, understand and handle abstract concepts, and use knowledge to manipulate one's environment". This essentially human-oriented definition could be modified in the context of nonhuman organisms as the ability to recognize and solve problems, learn from experience, and adapt to and modify the environment according to the requirements of a given individual or species. One example of the "intelligence" of plants and bees is that some plants infuse their nectar with low levels of nicotine and caffeine, and bees appear to prefer flower nectar containing low concentrations of caffeine. Caffeine has a beneficial effect on the long-term memory of bees, which enables them to remember the particular flowers that have caffeinated nectar and visit them more often than flowers with non-caffeinated nectar, thereby facilitating the pollination of the former (Chittka and Peng 2013). Thomson et al. (2015) also reported that nectar containing low levels of caffeine increases bumble bees' interaction with the flowers and improves pollination, although high caffeine levels lead to aversion. It may not be totally inappropriate to interpret this as an "intelligent" act by the plant.

Calvo and Baluška (2015) found that diverse groups of organisms may be observed to share "minimal forms of intelligence" across a large number of taxonomic groups. For example, organisms as diverse as *Drosophila* larvae, the nematode *Caenorhabditis elegans*, and roots of the model plant *Arabidopsis thaliana* and that of the maize *Zea mays* exhibit similar light-induced escape behavior. Both the worm *C. elegans* and the roots of the plant *Arabidopsis* produce hydrogen peroxide upon exposure to light. Many more examples may be cited where evolutionarily distant groups living in similar habitat have evolved similar behavioral strategies to adapt to a similar type of challenge or stress. Again, plant roots have a rich repertoire of behavioral responses such as moving downward into the soil, sensing and avoiding obstacles to chart a course, avoiding light, and exhibiting positive or negative response to certain chemicals, nutrients, or salt concentrations. Calvo and Baluška (2015) asked whether all this could be attributed to the "hard-wired", innate "instincts", or should we look into the possibilities of the role of intelligence and learning, even in the minimal sense of these terms. Calvo (2017), therefore, has examined the possible existence of plant subjectivity within the framework of the emerging discipline of plant neurobiology. Plant neurobiology tries to fathom the question of plant intelligence by probing plant signaling and behavior to understand how plants "perceive and act". In an evolutionary perspective, consciousness appears to have evolved during the "Cambrian explosion", which not only involved land mammals, but land plants as well. There is mounting evidence that plants can perceive the environment and shape their actions accordingly, despite the absence of a central nervous system. Plants generate APs, which are mediated by ion channels, just as in animals. However, in most animals, APs are transmitted along neurons, which the plants lack, though they use the phloem for transmission. This shows that a central nervous system is not a must for plants to be able to communicate and coordinate. Climbing parasitic plants have been observed to grow toward both tomato and wheat plants, but prefer the former, because the latter releases a repellant in its chemical package. It is possible that consciousness evolved multiple times during the evolutionary history, and might have involved different structures and strategies, not necessarily limited to the development of large brain size and a neocortex. Feelings evolved early in evolution, and involve the brainstem in animals. However, keeping feelings confined only to animals probably reflects the human tendency to zoomorphize or anthropomorphize behavior. Though plants lack neurons, J.C. Bose showed the existence of plant action potential which are transmitted via the phloem that serves as a "phytoneuron". These observations of Bose are now supported by many modern observations. Consequently, they at least keep the question of plant sentience open and deserving further study. Trewavas (2017) reiterated that the presence of brain, neurons, and a nervous system is not essential requisites for intelligent behavior. Plants can detect and respond to many environmental cues that enable them to access unevenly distributed resources and optimize their exploitation. They can also use volatile organic compounds (VOCs) for communication among individuals, which may serve as a "language" for recognizing "self and alien species". In a recent study published in the journal *Nature Communications*, Japanese scientist Yuri Aratani and his co-researchers showed that VOCs released by caterpillar-damaged plants elicited calcium ion signals from *Arabidopsis* plants in their vicinity. Thus, the VOCs served as "airborne signals", which could be received and interpreted by neighboring plants. This "interplant communication" served to protect plants from various "environmental threats" such as mechanical or herbivore-induced injury (Aratani et al. 2023). Both negative and positive (mutualistic) biotic interactions are known to exist between plants and their associated microbiome. Calvo et al. (2021) argued that the general contention that plants lack consciousness stands on the basis of human consciousness criteria, and therefore, constitutes a "top-down attitude". The belief that plants are not conscious entities is based on consciousness theories that involve the presence and state of nervous system. However, there are some consciousness theories that are not based on nervous system. The integrated information theory is one such theory that is not based on nervous system or brain, and tries to show that plants have "a minimal form of consciousness". They point out that the study of consciousness conventionally involves psychology, philosophy, and neurobiology. The benchmark fixed in neurobiology is essentially human, and is extrapolated in other mammals and birds. Even in these groups of animals, which are relatively close to the humans, this extrapolation is fraught with difficulties. It loses its tenability further in the more distantly related groups of organisms, more so in plants. Comparing behavioral elements among diverse groups of organisms is hardly meaningful, as behavior is dictated by evolution and various adaptive mechanisms in a given environment. The biological approach, on the contrary, is evolution-based and more bottom up. Stuart Sutherland gave the definition of consciousness, which is "having perceptions, thoughts, feelings; awareness. To be conscious it is only necessary to be aware of the external world" (Sutherland 1996, as cited in Calvo et al. 2021). In the sense of an "awareness of the external world", consciousness is present in Protozoa, which are thought to be the earliest eukaryotic cells. They can perceive diverse types of stimuli and are able to map their environment in order to optimize food capture and escape from adverse influences. Protozoa have also been shown to have conditioned or associated behavior (Calvo et al. 2021). However, Mallatt et al. (2021) have negated the applicability of integrated information theory (IIT) in explaining and justifying plant consciousness (Calvo et al. 2021). By assuming that all cells are conscious, Calvo, Baluška, and Trewavas have developed their premises of plant consciousness. One of the reasons for the rejection of IIT is that it is possible to confer consciousness to non-living systems within the premises of the IIT, which, therefore, is inappropriate and inadequate for explaining consciousness in biological organisms.

3.3.6 Do Plants Have Associated Learning?

There is an interesting ongoing debate on whether plants have associated learning or not. Monica Gagliano and her co-researchers (Gagliano et al. 2016) used a Y-maze to test pea seedlings, which were found to have the ability to develop an association of light with a neutral cue, and this learned behavior overrode the innate phototropism of the plants. Based on these results, Gagliano (2017) conjectured that the practice of taking neural structures as the prime determinant of cognitive ability needs to be reexamined. However, Markel (2020) reported that a repetition of the experiment with a larger sample size did not produce a similar result, calling for further confirmation of this sensational finding. In their response to Markel (2020), Gagliano et al. (2020) have pointed out that Markel's experimental setup did not satisfy relevant conditions. For example, while the original experiment was conducted inside a completely dark room with a distance of 20 cm between plants placed in Y-mazes, Markel's experiment was carried out in a smaller chamber which was not completely dark inside. Because of this, light from multiple sources resulted in random growth patterns. Markel (2020 a, b) responded by saying that the smaller size of the experimental chamber and more compact spacing could not be sufficient reasons for lack of replication of the results. He, however, did not dismiss the possibility of the existence of associative learning in plants, but called for more detailed experimentation to confirm the replicability of the results. Parise et al. (2020) defended the cognitive abilities of plants by stating that despite lacking brains and neurons, plants exhibit several complex behavioral traits that enable them to cope with their constantly changing environments. Plants can gather and process information, can learn and make decisions, and have memory, which testify for their cognitive abilities. Furthermore, the "extended brain hypothesis" advances the idea that cognition can extend beyond the confines of the body into the environment. Humans can transfer their ideas on a piece of paper or on devices like computers and smartphones to transfer an idea into "external devices". Spiders extend their cognitive capacities to their web. Similarly, plants may be able to "offload" their cognitive abilities through their root exudates and their associated microbiome, for example, the mycorrhizal fungi. Armed with this ability, plants can perform complex behavior without having brains and neurons. Loy et al. (2021) provided an overview of the extensive experiments conducted on associative learning in invertebrates, plants, and microorganisms, because of the importance given to associative learning for explaining the origin of consciousness. They found that there is lack of conclusive and convincing evidence for the presence of associative learning in plants, which calls for further studies.

3.3.7 Plant Consciousness: A Debated Issue

Coming back to the issue of plant consciousness, contrary to the claims of Trewavas (2017) and several other plant neurobiologists, Taiz et al. (2019) contended that the emergence of consciousness requires the brain to evolve a high degree of complexity in its structure and function. Even among the animals, only vertebrates including fish, and some invertebrate groups such as insects, crabs, and cephalopods like octopuses and squids, can be considered to have evolved consciousness. Therefore, plants do not possess the required level of organizational complexity and lack brain and neurons for developing consciousness, despite the claims made by plant neurobiologists. In a continuing debate, Calvo and Trewavas (2020), in their response to the contention that plants do not require or have consciousness (Taiz et al. 2019), pointed out that the article of Taiz and his co-authors adopted the "orthodox" approach which is based on an "animal-centric" definition of consciousness that dismisses the idea that plants can possess consciousness. The authors further stated that instead of setting some "zoo-centric" physiology-based benchmarks to dub plants as simple organisms, plant signaling and behavior need to be analyzed in information-processing terms to appreciate the complexity required for plants to survive in a fluctuating environment. Complementary models rather than conflicting ones, therefore, should be used to arrive at an integrated understanding. Taiz et al. (2020) replied to the plant neurobiologists by reiterating that instead of developing subjective consciousness as done by some animals, plants have gone along a different evolutionary course by evolving adaptive behavior determined by natural selection and environmental factors acting epigenetically. Such adaptive behaviors cannot be used as indicators of "consciousness and cognition". They further pointed out that "information processing" does not imply "consciousness, cognition, or volition". In their defense of the cognitive abilities of plants vis-à-vis their possession of consciousness, Segundo-Ortin and Calvo (2022) argued that plants have been shown to exhibit a number of cognitive capabilities earlier believed to be only present in animals. Therefore, if plants are considered even minimally cognitive, they should also be called at least minimally conscious. The cellular basis of consciousness (CBC) hypothesizes that consciousness is a fundamental property of all cellular life. Several subcellular structures can play significant roles in the emergence of consciousness. These include excitable membranes, microtubules and actin filaments making up the cytoskeleton, and structurally flexible proteins. All these arguments and counter-arguments keep the question of plant consciousness open and alive.

3.4 Conclusions

The information reviewed in the preceding sections suggests that it may be more prudent to look for sentience and consciousness much further down the evolutionary line than was earlier proposed. For example, the notion that only humans – or the primates and possibly all mammals – have the ability to feel pain may be dubbed as an anthropocentric-zoocentric fallacy. In the beginning, only humans were believed to have consciousness. Subsequent research led to the inclusion of nonhuman apes, then all primates, followed by all vertebrates including fish, and more recently, even some invertebrates. And finally, a lively debate is continuing on the question of plant consciousness. The facts presented in many of the publications reviewed in this chapter strengthen the conviction of moving away from an anthropocentric to a biocentric and even ecocentric environmental approach in studies on plant and animal behavior. The links between core and extended consciousness and the corresponding core and autobiographical self are indicative of the thread of continuity that runs between all life and the ecosystems in which they evolve.

4

"THE SHALLOW AND THE DEEP" ECOLOGIES

4.1 Introduction

Deep Ecology, which recognizes intrinsic value in all forms of life, harbors ecocentric ideals. The Norwegian philosopher Arne Naess (1912–2009) coined the term Deep Ecology and, while welcoming the growth of environmental awareness, was critical of the overt emphasis given by the contemporary ecologists on ameliorating the effects of pollution and resource depletion with emphasis on affluent and developed societies. This, Naess said, is the Shallow Ecology, which he distinguished from the Deep Ecology movement, and contended that the latter recognizes all species of organisms as components of the ecological web with "intrinsic relations" between each other. This "total field model", therefore, negates the "man-in-environment concept". The Deep Ecology movement also introduces the concept of "Biospherical egalitarianism – in principle" and recognizes that not only humans but every life-form enjoys an *"equal right to live and blossom"* (*italics* by the author), though "some killing, exploitation and suppression" are inevitable for practical purposes of survival. The Deep Ecology movement also promotes diversity – including human cultural diversity – co-existence, and cooperation, with the motto of "live and let live" being the foremost priority. "The principles of ecological egalitarianism and symbiosis" (Naess 1973, 95–96) implicit in the Deep Ecology movement are also against differences between developed and developing nations. Tackling the problems of pollution and unsustainable exploitation of resources through a more holistic approach and sustaining complexity (e.g., that in an ecosystem) but avoiding chaotic conditions are the other two approaches favored by the Deep Ecology movement, which also

DOI: 10.4324/9781003481058-4

advocates "local autonomy ... decentralization", local self-reliance, and ecological security. However, Naess further clarified that the principles of Deep Ecology are not to be viewed as strongly normative, but more as a "value priority system" (Naess 1973, 98–99).

4.2 Growth of Deep Ecology

George Sessions provided an elaborate overview of the emergence of the Deep Ecology movement and the background of its genesis, beginning with the ecocentric thoughts of Aldo Leopold found in his *A Sand County Almanac*, followed by the writing of *Silent Spring* by Rachel Carson in 1962, which can be said to have marked the beginning of "the Age of Ecology". However, the US government's approach to conservation – exemplified by the publication of the book *The Quiet Crisis* by the US Secretary of the Interior Stewart Udall in 1963 – adopted a distinctly anthropocentric approach, viewing the environment as an assemblage of "resources". The ecocentric underpinnings of the American conservation thought as evidenced in the writings of Henry David Thoreau, John Muir, and Aldo Leopold did not get due attention and were not adequately reflected in conservational policy making (Sessions 1987). Lynn White Jr. pointed out the anthropocentric treatment of nature in mainstream Christianity, which led to the prevalence of a philosophy of domination and exploitation of nature in Western culture. This, in turn, deeply influenced the directions taken by modern science and technology. Radical critiques of this anthropocentric bias had also come from ecologists and philosophers such as Raymond Dasmann, John Milton, Frank Egler, Paul Shepard, Eugene P. Odum, and others. Much of these radical inputs are based on Zen Buddhism, Hinduism, and Taoism, as well as on sound ecological principles. The origins of ecological thoughts were traced by Sessions to the Pantheism of Baruch Spinoza, the transcendentalism of Emerson, Thoreau, and Walt Whitman, the writings of George Parkins Marsh, John Stuart Mill, George Santayana, and many others from Europe, USA, Canada, Australia, and other parts of the world. It is reflected in the literary traditions of D.H. Lawrence, Aldous Huxley, T.S. Eliot, and a host of other poets and novelists. In this way, ecocentric, Deep Ecological thoughts have gradually formed a rich opposing tradition to that of the mainstream anthropocentric traits of environmentalism. Peter Zapffe developed "biosophy" in 1941, followed by the use of the term "ecophilosophy" by Sigmund Kvaloy in 1974. As said earlier, Deep Ecology gets its primary directions in Arne Naess's 1973 lecture, "The Shallow and the Deep, Long-Range Ecology Movements". The Deep Ecology movement negates all short-term, utilitarian objectives of environmental protection, which give too much emphasis on pollution. Furthermore, these short-term approaches ignore the long-standing "biospherical relationships" (Sessions 1987, 112).

On the other hand, the connectedness and interrelationships of ecosystems, the principle of "biospherical or ecological egalitarianism", and "biocentric equality" (Keller 1997) that impart equal rights to every species to live and flourish, a deeper understanding about the complex structure and functioning of ecosystems, and a "deep-seated respect and ... veneration for nature" are the hallmarks of Deep Ecology to make it represent a "paradigm shift" in the environmental protection and conservation initiatives worldwide. Deep Ecology poses a challenge to the dominant Western ethic about nature. However, Sessions maintained that Deep Ecology should not be misunderstood as having an objective of creating an ethical framework like that of the Western ethics, as was also suggested by Naess. Deep Ecology also aims to broaden its scope to admit more diverse groups of ecological thinkers and activists with different philosophical and religious backgrounds by framing the Deep Ecological platform. Though Deep Ecological ideals cannot be said to have found a wide-ranging acceptance with the policy makers, a Deep Ecological worldview was able to stand on a firm footing by the 1970s or 1980s. Since then, its different facets have been studied and elaborated by various exponents.

Bill Devall (Devall 2001) also gave a detailed overview of the Deep Ecology movement, which culminated in the "ecosophy" of Arne Naess. The "passage of the federal Endangered Species Act," which incorporates the ecocentric principle of not granting humans the "right to willfully cause the extinction of other species, regardless of their value, or lack of value, for humans," is cited as one of the major outcomes of the Deep Ecology and similar holistic movements (Devall 2001). The Deep Ecology movement also led to the formulation of the eight-point Deep Ecology platform, which is elaborated here.

4.3 Platform Principles of the Deep Ecology Movement

The following principles of Deep Ecology have been summarized in the Deep Ecology platform (Devall 2001; Drengson et al. 2011).

1. All living beings have intrinsic value.
2. The diversity and richness of life have intrinsic value.
3. Except to satisfy vital human needs, humankind does not have a right to reduce this diversity and richness.
4. It would be better for human beings if there were fewer of them, and much better for other living creatures.
5. Today the extent and nature of human interference in the various ecosystems is not sustainable, and lack of sustainability is rising.
6. Decisive improvement requires considerable change: social, economic, technological, and ideological.

7. An ideological change would essentially entail seeking a better quality of life rather than a raised standard of living.
8. Those who accept the aforementioned points are responsible for trying to contribute directly or indirectly to the realization of the necessary changes.

Deep Ecology influenced the "ecosophy" of Al Gore, although he expressed some criticisms of its principles as well. Deep Ecology principles also find their expression in the "World Charter for Nature" passed by the United Nations General Assembly in 1982. The Earth Charter launched in 2000 also incorporated the philosophy of Deep Ecology and ecocentrism in its principles. In spite of these widespread incorporations, Deep Ecology has so far had limited influence in public life and policy, though it has a broad outlook on the other forms of global crisis, such as the social and economic crisis, war, organized violence, and so on. Deep Ecology advocates the principles of nonviolent activism based on the principles of Henry David Thoreau, M.K. Gandhi, and Martin Luther King Jr. to raise a popular voice against the challenges of environmental and moral degradation and violence. It is possible to accept the idea of a broad and all-encompassing ecological self in both western and eastern cultures. Thus, Deep Ecology has a wide scope for global acceptance and incorporation into public policy and ways of life. In an earlier review of the Deep Ecology movement, George Sessions also described in detail the development of Deep Ecological thoughts, and noted that "an intellectual foundation for a Deep Ecological world view is now largely in place". He cited the emergence of several ecologically oriented philosophies and activisms, such as "ecofeminism, Green politics", and others, that set the stage for "another major deep ecology movement" (Sessions 1987, 121).

In a more recent review of Deep Ecology, Drengson et al. (2011) emphasized that the platform principles of Deep Ecology are operative at multiple levels: local households and community, state, and global levels. Having a broad scope, these can accommodate a diversity of cultures and philosophies. The principles and concepts that distinguish Deep Ecology from humanism and anthropocentric philosophies include biospheric egalitarianism, ecocentrism, and recognition of intrinsic value in nonhumans. Naess also defined ecosophy as "a philosophy of ecological harmony or equilibrium" (Drengson et al. 2011). Ecosophy is a broad philosophy, where each person can have his or her own version of ecosophy based on the local attributes or a particular ecosystem or landscape or environmental component that s/he is especially concerned with. For example, Naess's ecosophy is named Ecosophy T, where T stands for his hut Tvergastein, located in an arctic setting in the mountains of Norway.

Naess formulated the basic norms and hypotheses (N: norms; H: hypothesis) of Deep Ecology, which are:

N1: Self-realization.
H1: Higher self-realization enhances identification with others.
H2: When anyone attains a high level of self-realization, its further increase depends on the self-realization of others.
H3: Complete self-realization of one depends on that of all the others.
N2: Self-realization for all living beings! (Naess 1989, 197)

He also laid down some "norms and hypotheses" which have their origins and implications in ecology:

H4: Diversity of life increases Self-realization potentials.
N3: Diversity of life!"
H5: Complexity of life increases Self-realization potentials.
N4: Complexity!
H6: Life resources of the earth are limited."
H7: When resources are limited, "symbiosis maximizes Self-realization potentials."
N5: "Symbiosis!" (Naess 1989, 199)

The tenets of Naess's ecosophy also uphold the principles of decentralization and local autonomy:

H8: "Local self-sufficiency and cooperation increases "self-realization."
H9: Local autonomy enhances self-sufficiency.
H10: In contrary to H9, centralization reduces "local self-sufficiency and autonomy"
N6: "Local self-sufficiency and cooperation!
N7: Local autonomy!
N8: No centralization! (Naess 1989, 206)

Naess believed that "any global policy of ecological harmony" has to recognize the differences between "needs" and "wishes" "on the basis of a system of values" (Naess 1989, 206). He also probed deep into the theoretical underpinnings of Deep Ecology and traced the links between conservation and non-violence. He referred to Gandhi who professed his belief in "Advaita", that is non-duality and essential unity of humans and all living beings. This precept he (Gandhi) practiced in his own life, and instructed his followers not to harm even poisonous creatures like snakes, scorpions, and spiders. Naess also recounted the influence of Buddha on his ecosophy. Buddha's teachings require his disciples to embrace all living beings like a mother embraces her offspring, and in the same way "a human self could embrace all living things" (Naess 1987).

4.4 Gandhian Thoughts and Deep Ecology

Mohandas Karamchand Gandhi (1869–1948), the well-known Indian nationalist leader and social and political thinker, led the nonviolent non-cooperation movement against the British colonial rule in India. He also nurtured deeply ecocentric ideals which inspired Deep Ecology adherents, especially Arne Naess, the founder of the Deep Ecology movement. In many of his utterances and writings, Gandhi's convictions show an unflinching faith in ecocentrism and a feeling of altruism for the nonhuman world. Some of his well-known quotes are presented here, which reflect the ecological thoughts of this exponent of peace, non-violence, and love and respect for all forms of life. For example, in his *The Gospel of Sarvodaya*, he wrote: "I believe in the essential unity of man and, for that matter, of all that lives" (Prabhu and Rao 1967, 286). Again, in *Young Indian* (YI) of October 15, 1931 (Prabhu and Rao 1967, 309) he asserted: "I claim fellowship with the lowest of animals" (Prabhu and Rao 1967, 404). Gandhi's opposition to cow slaughter is well known, but his thoughts were based on the broad ideals of non-violence and ecocentrism, because he once said, "I would not kill a human being for protecting a cow, as I will not kill a cow for saving a human life, be it ever so precious" (YI, May 18, 1921) (Prabhu and Rao 1967, 529). In *An Autobiography*, he wrote:

> To my mind the life of a lamb is no less precious than that of a human being. I should be unwilling to take the life of a lamb for the sake of the human body. I hold that the more helpless a creature, the more entitled it is to protection by man from the cruelty of man.
> *(Prabhu and Rao 1967, 529)*

To Gandhi, cow protection held a much wider connotation, which was reflected in his statement that "In its finer and spiritual sense the term cow protection means the protection of every living creature" (Gandhi 1999, Vol. 30, 24). Gandhi's humanism was not confined to humans and extended beyond to include the entire living world. This is evident from several of his statements, such as

> my philosophy, my religion teaches me that brotherhood is not confined merely to the human species; that is, if we really have imbibed the spirit of brotherhood, it extends to the lower animals … how *(for)* a man who loved his fellow men it was obligatory to love his fellow-animals also, taking the word animals to mean the sub-human species.
> *(Gandhi 1999, Vol. 32, 244)*

and "My nationalism is as broad as the universe. It includes in its sweep even the lower animals. It includes in its sweep all the nations of the earth"

(Gandhi 1999, Vol. 32, 247). He penetrated to the core of Deep Ecological thinking, when he wrote in the October 25, 1925 issue of *Navajivan* (New Life: a magazine in Gujarati):

> This realization of the Self, or Self-knowledge, is not possible until one has achieved unity with all living beings – has become one with God. To accomplish such a unity implies deliberate sharing of the suffering of others and the eradication of such suffering.
>
> *(Gandhi 1999, Vol. 33, 154)*

He also said: "The truth is that my ethics not only permit me to claim but require me to own kinship with not merely the ape but the horse and the sheep, the lion and the leopard, the snake and the scorpion" (Gandhi 1999, Vol. 36, 5); and

> I intensely yearn to serve the animal world. …. Just as service to one's country …. is also service to humanity, so my service of human beings includes service to the animal-world. … my service to human beings is not incompatible with the welfare of the animal-world.
>
> *(Gandhi 1999, Vol. 31, 475)*

It is also worthwhile to note that Gandhi did not support any hypocritical ritualistic following of non-violence with an ulterior design, and had a word of caution for such mindsets by saying that

> It should be remembered too that mere *jivadaya* (kindness to animals) does not enable us to overcome the "six deadly enemies within us" … the *jivadaya* of a person who is steeped in anger and lust, but daily feeds the ants and insects and refrains from killing has hardly anything in it to recommend itself. It is a mechanical performance without any spiritual value. It may even be worse – a hypocritical screen for hiding the corruption within.
>
> *(Harijan September 15, 1940, 285; Prabhu and Rao 1967, 578)*

Gandhi longed for purification from within, and not in outward demonstrations of piety toward humans as well as nonhumans.

4.5 Heidegger's Philosophy and Deep Ecology

It has been suggested that the basic principles of Deep Ecology can be regarded as concordant with the philosophical thoughts of Martin Heidegger, the German philosopher. Heidegger's philosophy is often considered to have affinities with "Eastern traditions such as Vedanta, Buddhism, and Taoism". Zimmerman (1993) traced the harmony between Heidegger's thoughts and Mahayana Buddhism to their common traits of the rejection of anthropo-centrism and man-nature dualism, which are held responsible for the pre-sent-day environmental crisis.

Magdalena Holy-Lucsaz contends that an examination of the nature and extent of the support provided by Heidegger's philosophy to the "founda-tional assumptions" of Deep Ecology can be highly rewarding (Holy-Lucsaz 2015). One central issue around which this examination can be done is to look at the negation of the concept of the "great chain of being", whose ori-gins could be traced to the philosophy of Plato who wrote about "degrees of completeness". From this core concept, Aristotle developed his "scala natu-rae" or Ladder of Life model which is a hierarchical depiction of the vari-ous groups of organisms. At the bottom of this ladder is inanimate matter, stacked above which are the lower plants, followed by higher plants. Above the vegetable realm are the "Zoophyta", with sponges and jellyfishes at the lowest rung, above which the different animal groups are stacked in succes-sive layers: first come mollusks other than cephalopods, then arthropods other than crustaceans; above these are the crustaceans, then cephalopods, fish, amphibians, reptiles and birds, whales and dolphins, other mammals, and finally man occupying the apex. Above men are placed the angels and the creator. Plants have only "nutritive soul", all animals other than humans have "sensitive soul" (i.e., nutritive + sensitive souls), and humans have "rational soul" (i.e., nutritive + sensitive + rational souls). Thus, this arrangement is a "nested" one which implies that the higher categories have the souls possessed by the lower ones in addition to that possessed solely by them (Cohen 2004). This hierarchical scheme had important implications in recognition of rights to exist. This arrangement was later interpreted to give organisms occupying the higher levels more rights to exist, while beings at the lower levels were subordinate to those at the higher levels, and con-sequently had lesser rights than those at the higher levels, as was suggested by the German mathematician-philosopher Wilhelm Leibnitz (1646–1716) (Holy-Lucsaz 2015). However, it would also be pertinent to point out that despite this sort of an interpretation, Aristotle's hierarchical model ought not to be seen to have been framed with an exploitative or domineering attitude by its creator, but was simply meant to reflect the stages in develop-ment of the different groups of organisms. His analysis of the evolution of "soul" in different groups could be matched with the concepts of proto-self, core-self-core consciousness, and autobiographical self-extended conscious-ness as proposed by Antonio Damasio (Damasio 1998, 1999; see Chapter 3). Its subsequent interpretation from an anthropocentric, exploitative view of granting human dominance over the other life forms might be more respon-sible for its negation by the environmentalists. Although Aristotle had said that plants are for use of animals and animals for the sake of man, it is likely to be more in a hierarchical sense and not in an exploitative sense. Just as plants utilize inanimate matter such as minerals to carry out their life processes, and animals in turn utilize plants for their survival, growth, and reproduction, humans also depend upon animals (and plants) in the food chain to survive, flourish, and reproduce. Food chain is not exploitative but implies sustenance and survival. Therefore, distinction should be made between human sustenance and exploitation. The right for sustenance and prudent utilization is also recognized in the third platform principle of Deep Ecology, which talks about the use of nonhuman diversity just enough to satisfy vital human needs. A very important Aristotelian principle is the "principle of mean", which automatically forbids excess of anything and would logically exclude both abstinence as well as over-exploitation of the natural world. Aristotle also said that prudent utilization comprised the virtuous path, and that a virtuous man adhered to the mean and avoided both excess and deficiency (Hardie 1964–1965). Thus, virtue was achieved by maintaining the mean, which represented the balance between two extremes. This implies that since Aristotle was conceptually against excessive exploitation and utilization as a principle, it also should logically extend to the natural world and its myriad components. We can also argue that because Aristotle was the first to evolve and lay down a formal system of reasoning, the principle of mean ought to be applicable in all situations and contexts. Therefore, his statement that plants are for the sake of animals which in turn are for the sake of humans can be interpreted to imply sustenance – that is prudent use around a "mean" – and not unbridled exploitation as has been suggested later. Nevertheless, though Aristotle could perhaps be defended for his hierarchical model of life, radical ecological philosophies like Deep Ecology are opposed to any overt hierarchical treatment of the living world. The first two "platform principles" of Deep Ecology declare that all life on the Earth have intrinsic value and that the "diversity and richness" of life are also intrinsically valuable. Therefore, if all living beings are intrinsically valuable, then there is no scope for recognizing a hierarchical scheme of rights, which goes against the principle of "ecospherical egalitarianism" – a principle that Warwick Fox called a defining term for the ethical position adopted by Deep Ecology (Fox 1995, 117–118, as cited in Holy-Lucsaz 2015). The intellectual support for this stand of Deep Ecology may be found in the philosophy of Heidegger, especially in his works after 1935. Holy-Lucsaz shows that the western ontotheology asks the question as to what is the "supreme being", and this implies that the supreme being has the highest degree of "being-ness", with the other beings representing a "graduality" of the degree of beingness. This concept of gradation agrees well with the hierarchical structure of the great chain of being. Heidegger's thoughts repudiated this gradation and negated the presence of beingness in terms of degrees. Holy-Lucsaz

(2015) – citing Richerson (2003) – argued that contrary to the postulates of anthropocentric ontology, t/here-beings (humans) do not create beings, but are embedded in a vastness of beings. This transformation of relation between humans (t/here-being) and other beings brings about a realization that all beings are equal in Heidegger's philosophy (Gray 1957, 197–207, as cited in Holy-Lucsaz 2015).

Zimmerman (2003) explored the relevance of Heidegger's phenomenology in environmental philosophy. Heidegger was critical of anthropocentrism, though he was not a biocentrist in its strict sense, because he believed in human discontinuity from nature, based on the findings in "physics, chemistry, biology, and psychology". Heidegger did not examine the issue of man-nature relationship with an axiological (value-based or value-oriented) approach. His approach was more ontological (based on the nature of being) – based on the premise that the interactions of t/here-being, that is humans with the other beings took place in the "temporal-historical clearing opened up through Dasein" (German: literally, being there or existence: the existence of the human individual). This clearing has neither been created by Dasein, nor do the humans become masters of this space. Humans are required to "let things be", so that all beings can express and manifest themselves in their different forms of "intelligibility". However, Heidegger started adopting this position in his later (post-1935) works, which is termed as the "turn" ("kehre") in his evaluation of humans vis-à-vis nature (Holy-Lucsaz 2015).

A few decades ago, Gray (1957) had also drawn attention to the change in Heidegger's perceptions about nonhumans in his later works. In Heidegger's understanding, humans are not separable from the world in which they are placed. Therefore, human existence can be found in the world around him, though Heidegger recognized that humans are distinct from the other beings and things in the world. Heidegger also developed the concept that the things of nature should be allowed to retain their true nature, and given freedom to exist. It follows that Heidegger's ontology recognized intrinsic value in all natural entities on the basis of the concept of Heidegger's "Das ding" or "This thing", where he distinguished between a thing and an object, a difference that reflects the prioritized objective of modern humans to commoditize nature by treating its components as mere resources over which humans must establish their domination (Howe 1993). If we examine Heidegger's ontology in the light of the first two platform principles of Deep Ecology, then the "let beings be" notion of Heidegger implies support for the first platform principle which states that both human and nonhuman life on Earth have intrinsic value, and that the well-being and flourishing of nonhumans are independent of their usefulness to the humans. The second principle stipulates that it is essential to maintain the "richness and diversity" of nonhuman life forms because these "contribute to the realization of these values and are also values in themselves" (Keller 2009). Naess (1997) spoke about the influence of the concept of non-dualism found in Mahayana Buddhism and Advaita Vedanta vis-à-vis the interest of Heidegger in Western Buddhism, Buddhism, and Taoism. This sharing of non-dualism is another common thread between Heidegger's thought and Deep Ecology.

4.6 Deep Ecology, Ecological Consciousness, and Ecological Resistance

Gandhi's concept of "self-realization", which was briefly discussed in Section 4.2, is not only achieved through a feeling of unity with and empathy for all forms of life, but it also calls for action to alleviate the suffering of not only humans but all forms of life. This is in consonance with the objectives of Deep Ecology. Bill Devall and William Devall also suggested that Deep Ecology is not only concerned with developing an "Ecological Consciousness", but it leads to "ecological resisting", which represents transforming "Ecological Consciousness" into "political action", with special reference to industrial societies like the USA (Devall and Devall 1982). This also brings into light the possible relationship between technology and Deep Ecology. Shallow Ecology approaches, which are aimed at reducing or ameliorating the ill effects of technology on the environment, are dependent on new and "clean" technological solutions to these problems. This approach mostly leads to the introduction of new ameliorative technologies, which, however, give only piecemeal solutions, and a string of technologies follow. Though each of these technological innovations promise a better and more complete solution, they produce a succession of new issues that often overshadow the actual achievements. This dependence on technology instead of relying on social, cultural, and even psychological reorientations is a major feature of the shallow ecological approach. For example, shallow ecology would rely on the use of scrubber stacks, which can lead to reduction of pollution from coal-fired power plants, but cannot address the question of changing the energy source to a more benign one, which needs societal reorientation toward the demands and preferences for energy. Thus, an attitudinal change is necessary, not just short-term reformist solutions. The former approach can only develop if the "dominant technocratic mentality" changes from visualizing humans as "lords and masters of the Planet" to co-existing partners with the natural components in the ecosphere. The modern society places technology at the central position, and not "religious and family institutions" as in "many so-called primitive societies". The concept of "Ecological Consciousness" can provide an answer to the environmental crisis, because it makes its bearer aware and conscious of the worth of not only humans, but that of nonhuman entities such as "rocks, wolves, trees, and rivers". This position is in agreement with Aldo Leopold's statement that whatever is right for the biotic community represents a good ethic, and that humanity's position is not to tower above nature, but to exist within the folds of nature. Thus, the spread of ecological consciousness in the society represents its approval of "resisting modernism". This philosophical view also links consciousness and conscience. Having this conscience can lead to "Ecological resisting", which opposes all short-sighted, piecemeal, and solely technological solutions to most environmental problems, including that of pollution. Such technocentric attitudes refuse to recognize and respect the vulnerability of many ecosystems such as rain forests, coral reefs, mountains, and others, and continue to push for technological solutions to the problems caused by unsustainable developmental activities in such areas. This attitude will only be abandoned when large sections of the society begin to grow ecological consciousness. Ecological resistance – which arises from a "shift in consciousness" – is totally nonviolent in nature, with its primary concern stemming from respect for the "intrinsic values of the biosphere", and not out of concern for human health and safety alone. An important goal of ecological resisting is to develop friendship with other entities such as nonhuman organisms, or entire ecosystems. An ecological resistance activist takes up the case for a beleaguered species or ecosystem and advocates for its protection. It is based on the Aristotelian concept of friendship, which involves working for "other's good for the other's own sake". That means such a friendship recognizes intrinsic value in the other, which here includes nonhuman species or ecosystems. The deep implication of this concept is that the self of the resister is extended to include others – a position that is adopted based on the principles of Gandhi who said that he was working for the good of a village not for any altruistic or humanitarian reasons, but for serving himself, because the self of each villager was a part of his own extended self. Though Deep Ecological principles can be said to advocate for "small scale, elegantly simple, nonviolent, understandable by laymen, flexible, and conservative" actions, all of these can be distorted and conveniently interpreted as "appropriate" to fit the dominant anthropocentric worldview. Therefore, the spread of ecological consciousness (and ecological resisting) in human societies is a prerequisite for ensuring sustainable co-existence of all living beings (Devall and Devall 1982).

The notion that the scope of Deep Ecology extends beyond the confines of an ethic, as asserted by George Sessions, persists in the more recent writings also. Keller (2009) identified Deep Ecology as a very forceful critique of western anthropocentrism, which has inserted its roots deep into the domains of environmental policy and action, and of environmentalism itself. Placing humans over all other living beings is a long-running tradition in western thoughts, and the biocentric egalitarianism of Deep Ecology negates this human hegemony. Various Deep Ecology thinkers have denied the existence of any boundary between human and nonhuman realms, and this realization leads to an outward extension of the boundaries of self to embrace more and more beings, thereby visualizing the existence of only "one big Self". Thus, this oneness with the living world also implies that when humans harm the biosphere or any part of the living realm, they actually cause harm to themselves also. The quest of Deep Ecology, therefore, is not for an environmental ethic, but for a larger, all-embracing "ecological consciousness" (Keller 2009, 207).

4.7 Critiques of Deep Ecology

Among the critics of Deep Ecology, Murray Bookchin stands out for his vehement, persistent critique of Deep Ecology. Bookchin (1987a, b) made a scathing critique of Deep Ecology where he branded it as "a vague, formless, often self-contradictory, and invertebrate thing ... spiced with homilies from Taoism, Buddhism, spiritualism, reborn Christianity, and in some cases eco-fascism". This trend was opposed to "a long-developing, coherent, and socially oriented body of ideas that can best be called social ecology" (Bookchin 1987a). Bookchin (1987a) also hurled disdainful epithets on Arne Naess and other Deep Ecology proponents, and several ecologists like Barry Commoner and Paul Ehrlich. However, one of the major issues around which the Deep Ecology-social ecology debate is centered is that the latter regards the exploitation of the environment to have its roots in the human exploitation of other humans. Bookchin criticized what he called the attempt of Deep Ecology to label "humanity" essentially as an ugly "anthropocentric thing". The emphasis of Deep Ecology (and other ecocentric worldviews) on getting the human society rid of its anthropocentric bias coupled with its association with traditional faiths and concepts such as animisms, Taoism, Buddhism, Hindu philosophy, etc. led Bookchin to brand Deep Ecology as an eclectic "hodgepodge" and a "spiritual Eco-la-la" (Bookchin 1987a). Keller (1997) also wrote in his critique that it was difficult to characterize Deep Ecology because of its "eclectic diversity" by virtue of its connections with a host of eastern religions and philosophies, and modern Western philosophers, poets (and other writers), artists, and environmentalists. He further said that it is difficult to identify the area of convergence of these myriad thought processes in an ecological context. However, what is branded as an "eclectic medley" could also be interpreted as integrative. To take such a position may be the preferred approach toward addressing an issue like the environmental crisis, which is a complex of scientific, technological, social, and very importantly, ethical-philosophical crisis. If the "proof of the cake is in eating", then Daoism, Buddhism, nature religions, and similar faith systems could maintain largely sustainable societies all over the world over a sufficiently long period of time without receiving any considerable support from technology. Therefore, it may be more prudent to search for some enduring principles within their domains rather than summarily dismissing them and branding them with unpalatable adjectives. Another major aspect of the Deep Ecology – social ecology debate that is especially relevant in the context of this book on ecocentric thoughts – is whether ecology is to recognize intrinsic value in nature and preserve it on that basis, or its aim is to create an ecological thought system based entirely on instrumental values of nature. It also needs to be clarified here that despite his opposition to Deep Ecology, Bookchin did not subscribe to the human-engineered, instrumental vision of nature and was against large-scale interferences with natural processes. In fact, Bookchin and most other social ecologists believed in the co-existence of both human-modified landscapes as well as wilderness in an ecologically sustainable Earth (Chase 1991). Keller (1997) was more specific in his critique where he pointed out the inconsistencies of two major cornerstones of Deep Ecology, namely, "biocentric equality" and "metaphysical holism". The former assigns intrinsic value to all living beings and does not recognize any gradation in this value. It also does not attach any priority to humans over nonhumans. Keller contended that such egalitarianism will not work in case of "conflicting interests". Therefore, for any environmental ethic to be effective, it has to recognize differential values among species and individual organisms. The second principle of metaphysical holism advocates abolishing "the ontological boundaries between self and other … through the process of self-realization". Keller terms this form of metaphysical holism as "expansionary holism", which recognizes "only one big Self, the life world". Keller cited several examples to show that expansionary holism would fail to resolve a conflict between the members of this "larger whole". He also pointed out the inconsistencies between biocentric equality and metaphysical holism, because believing in or adopting one would automatically negate the other.

Arne Naess gave a pacifist response to the aggressive critique of Bookchin during his correspondence with John Clark during 1988 to 1997 (Clark 2010). In one of his letters to Clark, Naess responded objectively and optimistically to Bookchin's concept of social ecology, which states that the human inclination to dominate over nature stems from the tendency of humans to exploit other humans. Naess did not reject this viewpoint, but also pointed out that the framework for this concept emerged from the philosophy of Hegel and Kropotkin, which in his assessment was "excellent", and yet there was scope for accommodating other viewpoints. As Clark recounted, Naess said that he was "completely relaxed about social ecology/deep ecology" and was confident that the relation between the two disciplines would turn out to be a satisfying one. Nevertheless, he visualized that social ecology extended beyond the constricted framework conceived by Bookchin and included other thinkers who did not subscribe to the Hegel-Kropotkin framework. In other words, social ecology was much more inclusive than was perceived by Bookchin. In a further analysis of Bookchin's views, one may find that they do not show any path out of the dire predicament in which the reckless human exploitation of the biosphere accompanied by a scant regard for other life forms and ecosystems has landed us. Today, when the endless series of negotiations among world leaders fail to achieve any substantial progress toward checking the devastating effects of climate change, and when poor and vulnerable nations are threatened by engulfing seas and other natural disasters despite not being responsible for the greenhouse gas accumulation in the atmosphere, Bookchin's mocking criticism of the Deep Ecology precept of expanding the narrow, anthropocentric self into a wider cosmic 'S'elf sounds hollow. It needs to be mentioned here that Bookchin adopted a revised and much more moderate and reconciliatory stance a few years later in his "dialogue" with Dave Foreman (Bookchin and Foreman 1991) where he asked the ecologists to be careful while talking about the problem of overpopulation and ensure that such a move did not lead to coercive measures that could often turn racist and discriminatory. His second concern was on the question of safeguarding the natural world from the environmental degradation and destruction unleashed by human societies. Here, Deep Ecology tended to blame the human species as a whole instead of the role played by its specifically anthropocentric members such as the corporates. Bookchin expressed his apprehension that such clumping together of all sections of humanity played down the exploitation of women by men, underprivileged by the privileged, and workers by employers.

Clark (1996) also referred to Arne Naess's views in the context of having the Deep Ecology platform as a means to "facilitate cooperation" between all philosophies that adhere to ecocentrism. Thus, Deep Ecology also has the width of scope to accommodate diverse positions on specific aspects within the broad ambit of the philosophy of ecocentrism. Naess inducted the perspective of Gandhi that the differences of opinion on specific issues should not be carried down to cause conflicts among those holding opposing views. In fact, diverse views such as Deep Ecology, social ecology, and ecofeminism could further enrich the ecocentric domain. They are, therefore, not to be viewed as "mutually exclusive alternatives", but as different means with the basic objective of guiding the society's progress in an ecologically viable direction. Naess advocated that following the Gandhian principles of avoiding "mistrust and misjudgment", the proponents of the diverse ecocentric ideologies needed to focus on their commonalities and strive for an integration of their principles. Such a unified approach should be the hallmark of ecocentrism and Deep Ecology. Clark observed that these "norms" advanced by Naess ensured a "non-defensive, non-dogmatic approach". He further pointed out that Naess termed Deep Ecology as a "movement" and not a specific "philosophy" or "ideology". He and some other scholars suggested that Naess kept the Deep Ecology statements at a generalized level, and allowed a certain degree of "vagueness" in its "concepts and principles". This enables the Deep Ecology platform to accommodate a diverse set of ideas in its midst (Clark 1996). We can also see that the first platform principle states that both human and nonhuman life on Earth have intrinsic value, and that the well-being and flourishing of nonhumans are independent of their usefulness to humans. The second principle stipulates that it is essential to maintain the "richness and diversity" of nonhuman life forms because these "contribute to the realization of these values and are also values in themselves" (Keller 2009). Thus, it can be said that recognition of an ecocentric view of the world as a whole is a distinguishing feature of Deep Ecology. Keller (1997) – despite his critique of some of the Deep Ecology principles – also admitted that Deep Ecology had made lasting and valuable contributions to "the repudiation of the mechanical view of nature" espoused by western anthropocentrism. Deep Ecology has also driven home the realization that nonhumans including ecosystems have values higher than the mere use value attributed to them by humans. However, Keller concluded by saying that because of the inconsistencies mentioned earlier, Deep Ecology had failed to become an effective environmental ethic.

4.8 Conclusions

It is true that, even if the vagueness of Deep Ecology was deliberately maintained by its major proponents, it admittedly creates some problems on issues such as the right of humans to exploit the natural diversity to satisfy their "vital needs". What is problematic is that the term vital needs may vary from country to country or community to community, and could serve as a pretext to overexploit the nonhuman diversity. However, Naess always maintained that Deep Ecology is a movement which is a living entity that can revise and renew to enrich itself by accommodating new ideas. Deep Ecology's success, therefore, lies in its ingrained flexibility and accommodativeness. At this juncture, it may be worthwhile to remember that though philosophies and movements that profess ecocentrism such as Deep Ecology and others have evolved in more recent times, ecocentric perceptions of nature including both animate and inanimate nonhuman entities have been prevalent in human communities for most part of our history on this planet. Chapters 5 and 6 give a brief overview of this legacy that we have inherited.

5

INDIGENOUS VIEWS OF NATURE AND RECOGNITION OF ITS "SOUL"

5.1 Introduction

The philosophical debate on whether nature including its nonhuman living and even non-living components has intrinsic value began only lately, since around the 15th century. On the other hand, though not formally stated, recognition of intrinsic value of nature is implicit in many ancient texts, religious practices, worldviews, myths, and folklores of various tribes and other communities all over the world, a short account of which is provided in this chapter.

5.2 Human Evolution and Migration

Most scientists now agree that the first modern humans (*Homo sapiens*) originated in Africa ~300,000 years before present (BP). Humans evolved from this common stock and migrated to populate the whole world. Some early small-scale migration out of Africa might have taken place as early as 90,000–100,000 years BP, though these early migrants might not have survived for a period long enough to colonize other areas. The major wave of migration out of Africa is likely to have started around 80,000 years BP. Asia was populated first between 80,000 and 60,000 years ago. *Homo sapiens* tools that are similar to those found in Africa during the same period have been unearthed in Jwalapuram, Andhra Pradesh, India, that date back to 74,000 years BP. By 45,000 years BP or even earlier than that, modern humans had colonized Papua New Guinea and Australia. Their entry into Europe took place about 40,000 years BP, where they co-existed with the Neanderthals. The Americas were reached about 15,000 years BP (Gugliotta 2008).

DOI: 10.4324/9781003481058-5

Since their evolution about 200,000–300,000 years BP till about 12,000 years BP when the first cultivation started in the fertile crescent of Tigris-Euphrates valley in Mesopotamia, humans remained as "hunter-gatherers", subsisting on fishing, gathering fruits, tubers and other edibles, and hunting.

5.3 Nature in Hunter-Gatherers

How did the hunter-gatherers view nature which was all around them in its pristine form, and which gave them light, warmth, food, shelter, clothes, dyes, materials for making utensils and ornaments, and other means of sustenance? Did they worship nature and was it a part of their religion? Whether our hunter-gatherer ancestors had a religion – *sensu stricto* – still remains the subject of learned debates among researchers in this subject field. For the purpose of this book, we can simply try to at least get a glimpse of the perceptions in the minds of these early humans about the natural world, which with its myriad living creatures and non-living things surrounded them and whom they commonly encountered. Unfortunately, neither fossil records or old artifacts nor DNA samples provide any direct evidence of the perceptions of nature prevalent at a given time, though they may provide some indirect clues.

One way of understanding the perceptions of nature in the ancient hunter-gatherer societies could be to study their modern-day successors, who constitute a miniscule fraction of the several billions that live on the Earth today.

5.4 Animism(s) and the Recognition of Intrinsic Value in Nature

The Merriam-Webster dictionary defines animism as "attribution of conscious life to objects in and phenomena of nature or to inanimate objects". One of the definitions of animism in the Oxford Advanced Learner's Dictionary (Seventh Edition: 2005) is "the belief that plants, objects and natural things such as the weather have a living soul" or "the attribution of a living soul to plants, inanimate objects, and natural phenomena" (Google's Dictionary provided by Oxford Languages). Thus, animism recognizes intrinsic value in entities other than humans.

Early anthropologists like Edward B. Tylor (1832–1917) and James G. Frazer (1854–1941) thought of animism as a "magical philosophy" that tried to explain the workings of nature, albeit in a wrong way, because the adherents of animism did not possess proper scientific knowledge. Because of this ignorance, primitive men imputed souls in nonhuman entities in their quest for explaining the workings of nature. According to Tylor, animism was the most primitive religion that was based on illusion (Willerslev 2013). Tylor surmised that in animism, the spirits of nature are "modeled" after the concept of the human soul, and "explain nature on the primitive childlike theory that it is truly and throughout animated nature" (Tylor 1871, 270).

Primitive humans lived with the spirits of their ancestors, and "with the spirits of the stream and grove, plain and mountain", the sun radiating light and warmth, and the mighty sea. Just as humans lived and acted through their souls, souls present in other entities carried out the "operations of the world". Animism, which began as "a philosophy of human life", grew larger in its scope and extent to become "a philosophy of nature at large" (Tylor 1871, 271). For example, the Iroquois of North America thank the trees, shrubs, other plants, ecosystems such as springs, streams, and natural forces like wind, rain, and heavenly bodies such as sun, moon, and stars, which are needed by them to meet their various requirements. They appear to attach "real personality" to the spirits that animate the "world around them". Tribes of the African Gold Coast recognize a spirit or "*wong*" in trees, rivers, lakes, springs, termite hills, animals such as snakes, birds, crocodiles, elephants, and apes, or they may also think of these entities as *wong* themselves (Tylor 1871, 291). Similarly, volcanoes, rocks jutting out in rivers, and other objects could all appear animated to primitive humans. The Greeks inherited this notion of nature spirits, and subsequently, with their penetrating insight and reason transformed the animistic scheme to "physical science" to influence the whole world (Tylor 1871, 292–293). Tylor also noted that thoughts of this nature persisted even in modern humans. In certain moments of life, when people return to their childhood fancies, the stream appears to have a life like that of humans, when it cascades down "the hillside like a child". In the flow of water, "the poet's fancy can discern its personality of life". Thus, "the worship of well and lake, brook and river" teaches us that "what is poetry to us was philosophy to early man". Primitive humans thought that the movement of water was not governed by physical laws, but by "life and will" (Tylor 1871, 294–295).

Tylor quoted a Scottish rhyme composed by an anonymous poet, where two rivers assume human-like quality when they talk to each other:

> Tweed said to Till,
> "What gars ye rin sae still?"
> Till said to Tweed,
> "Though ye rin wi" speed,
> And I rin slaw,
> Yet where ye droun ae man,
> I droun twa!
>
> *(Tylor 1871, 295)*

Here, like two friends engaged in a friendly banter, the Till boasts that though relatively sluggish, it drowns twice as many people as the faster Tweed.

Another early and "legendary" anthropologist James G. Frazer is known all over the world for his famed publication *The Golden Bough* (Frazer

1890). However, Frazer's analysis and interpretation of ancient rites, rituals, myths, and folklores have been severely criticized by later anthropologists. He was accused of not giving a "plain account", and treated the rites or dogmas "*out of context*" (Strathern 1987). Frazer has also been criticized for regarding magic as primitive and wrong science (Kumar 2016). Frazer's contributions may be assessed without subscribing to his "evolutionism", which assumes that all societies pass through the same stages of development toward "intellectual and moral progress". Modern ethnographic studies negate this assumption and find them "not simply erroneous but frankly absurd" (Willerslev 2011).

5.5 Anthropological Approaches toward Animism and Totemism

Willerslev (2013) summarized the different anthropological approaches toward animism and totemism. According to him, early anthropologists like Tylor and Frazer thought of animism as a "magical philosophy" that tried to explain the workings of nature, albeit in a wrong way, because of the lack of scientific knowledge. Because of this fallacy, primitive men imputed souls in nonhuman entities in their quest to explain the workings of nature. Animism is the most primitive religion that is based on this illusion.

In the Durkheimian tradition, animism is symbolic. For example, many indigenous tribes regard the forest as their parents, which meets all their needs. Thus, the indigenous worldview is based on metaphors based on their kin relation such as that exists within a family. This reduces animism to a "false epistemology" based on a flawed philosophy.

The Lévi-Straussian view of structuralism considers totemism as a means of classification used by primitive humans, where particular clans/individuals are identified with a particular plant/animal/landscape element. Lévi-Strauss applied the principles of Saussurian linguistics to totemism, where each totem is like a linguistic sign to signify some entity. Thus, it is a classificatory system to distinguish, for example, the bear clan from the tiger or the turtle or some other clan (Willerslev 2013).

Descola (1996) talked about the "modes of identification" of "the boundaries between self and otherness", where totemism and animism are two such systems or modes of identification of humans and nonhumans. The totemic system is based on the "discontinuities between natural species" to construct a social hierarchical order, and has a classificatory approach. On the other hand, animism recognizes human properties and social elements in nonhumans with the objective of establishing relations between humans and the natural world. Totemism regards nonhumans as emblems or symbols called totems, which are the identifying marks of a clan or sometimes an individual. On the other hand, animism views them as relations. Descola (1996) highlighted the existence of two major modes of relation: reciprocity and predation. For example, the Tukanoan Indians of eastern Colombia in the Upper Amazon believe that internal exchanges between humans and nonhumans lead to the transformation of human souls into game animals, and this exchange maintains an equilibrium in the cosmos. For every game animal killed, a human soul is transformed to replenish the diminishing stock. In the predation mode, no such exchange takes place. Instead, game animals take revenge by sucking the blood of humans, and the "Masters of Animals" punish hunters who kill excessively by snakebites. In this way, reciprocal predation maintains a balance and regulates relations between humans and nonhumans. In a revised stand, Descola spoke of "ontological realities" where indigenous relations with the environment emanated from their everyday experiences with the nonhuman entities in their environment (Descola 2011, as cited in Willerslev 2013).

Viveiros De Castro (1998) examined animism against the concept of Amerindian "perspectivism" in which humans, animals, and spirits see each other in their own perspectives. In this, just as humans see themselves as humans, animals also see themselves as humans. The outer appearance of animals is akin to an "envelope" or "clothing" with an "internal human form" which is the "soul or spirit of the animal". Transformation and metamorphosis are the hallmarks of the world visualized in animism, where one animal can transform into another animal or a human, and humans can be transformed into animals. Many Amerindian cosmologies visualize a world where there is no essential difference between humans and animals, and where mutual transformations are the rule.

Tim Ingold (Ingold 2000, as cited in Insoll 2011) is of the opinion that animism (or totemism) is not "coherent and explicitly articulated doctrinal systems" but "rather orientations that are deeply embedded in everyday practice". Animism is perhaps better expressed as animisms, as there are both similarities and differences between the worldviews of different animistic communities. In the perspectives of animism, beings of all kinds – human and nonhuman and animate and inanimate – "continually and reciprocally bring one another into existence". Thus, it is much more than simply humans ascribing life to inert things, but where all beings generate one another and have viewpoints regarding the others. Willerslev termed this approach of Tim Ingold as counter-structuralism whose approach is phenomenological, where the natural world and humans are ontologically inseparable. This concept owes to the idea of "being-in-the-world" of Martin Heidegger. In fact, the analysis of Ingold pertains not only to animism, but generally to how humans perceive their environment. Therefore, animistic cosmology is practically oriented and is formed from different relational activities between humans and nature, such as during hunting, harvesting of fruits or tubers in particular seasons (that is in different contexts), or even in dreams (Willerslev 2013). Furthermore, totemism flourishes where social relations are of a vertical hierarchical structure. Animism, on the other hand, is horizontal and maintains an egalitarian relation with other entities. Hence, there is "extra-human perspectivism" in animism, whereas totemism is characterized by "a perspectivism" based on "inter-human metamorphosis" (Ingold 2000, and Pedersen 2001, as cited in Willerslev and Ulturgasheva 2012).

5.6 Some Case Studies of Animism and Totemism

In the Eveny (singular: Even) of Northeastern Siberia, the *khavek* – a guardian reindeer – protects the "open" and vulnerable bodies of children, which can be taken away by the ancestor spirits. The *khavek* disguises the human identity of the child to make it appear like a reindeer to deceive the malevolent spirit that attacks the child. Thus, there is a reversibility between the child and the reindeer. Furthermore, the reindeer serves as an intermediary between the human and nonhuman realms. The authors used Alfred Gell's distributed personhood concept (Gell 1998) to view the child as an extension of the guardian reindeer and vice versa. The Eveny regard the guardian reindeer as having "double soul". The child is the reindeer and the reindeer is the child (Willerslev and Ulturgasheva 2012).

To the pastoralist Darxad Mongols of the Mongolian Steppe, the mountains have "spirit-masters" (*ezed*) for whom people show deep respect. However, Humphrey was of the opinion that it is not "soul" that these entities contain, but "its own kind of indeterminate energy" which is present in "everything". This is "correlated with what a particular entity looks like". People recognize this "inner or concealed power of entities in the world", and can establish "intentional relations with them", as one can have with humans (Humphrey 1996, as cited in Pedersen 2001). This also perhaps shows the "inadequacy" of the western concept of soul in an eastern perspective, for example, the Indian concept of *atma*, which has a much broader perspective. It will be logical to conclude from the statement of Humphrey that North Asian animism does not comprise a "homogenous belief in nonhuman souls", but rather "heterogenous perceptions" of some sort of "interior spiritual quality in things". Thus, "the North Asia social realm is made up by both human and non-human beings". However, despite allowing in principle for a limitless socialization between the human and nonhuman worlds, thereby making them a boundless whole, the presence of many "asocial entities", which do not have any common social ground, results in pockets of discontinuity within this whole. For example, the human hunter and an animal predator like a bear have social relations because they share the same hunting ground. But in that sense, there are many beings among whom there is no common aspect, and therefore, they exist like the "holes in a Swiss cheese". Consequently, nature in North Asian animism does not exist as a "unified and unifying whole", but is governed by a "supersociality" that connects humans, nonhumans, and spirits, thereby giving the appearance of "a whole with holes in it" (Pedersen 2001).

Animism involves "analogous identification" within horizontal social relations which implies that human and nonhuman entities can be interchanged with one another, thereby providing a continuity between human and nonhuman domains. In an animistic worldview, every element can identify with another element, without any distinction between humans and nonhumans. This also engenders a "primordial sympathetic relationship" between humans and nonhumans. For example, the Koryaks of far eastern Siberia believe that any animal, such as bear, wolf, ermine, moose, and birds such as raven, can take their skins off and become humans, while the reverse is also true. A human can take on the skin of the bear or moose and become that animal. This transformation can take place between animals and birds also. A raven can put on a bear skin and get transformed into a bear.

The reindeer-breeding Psaatang in N. Mongolia strike a deal with the bear. If a bear, fleeing from Psaatang hunters, climbs up a Siberian larch (*Larix sibirica*) tree, then the hunters should not kill the bear, and walk away to allow the bear to escape. Similarly, if a Psaatang roaming alone and unarmed meets a bear and can climb up a larch, the bear is expected to reciprocate the non-aggression deal.

Among the Darxad Mongols, if a hunter meets a mountain antelope, and if the hunter has seen the antelope first, then he is free to shoot and kill the animal. However, if the antelope has spotted him first, then it is regarded as one possessing magical power and the hunter must be able to kill it with a single clean shot. If he fails, then he must get a shaman or a monk to remove the power transferred by the antelope to him (Pedersen 2001).

The works of early anthropologists like Tylor and Frazer have been the subjects of long and intricate debates, marked by initial outright rejection to be followed by more realistic reevaluations in the present context. The compilations of these early anthropologists along with the works of successive anthropologists are of great value in the context of the origin and evolution of ecocentric ideals in their different forms, and the recognition of extrinsic and intrinsic values in nonhuman entities.

A detailed study on the indigenous perceptions about the reindeer throws light on the nature of animism (Vitebsky and Alekseyev 2015). Reindeer (*Rangifer tarandus*) exists in both wild and domesticated states. Its domestication began ~2000–3000 years ago in southern Siberia, probably by the ancestors of today's Evenki and Eveny peoples. Domesticated reindeer are mainly used for pulling sledges, for saddling and riding in some areas. It is also milked in Siberia, and its meat consumed. Some are exclusively used for breeding. People have been living in close contact with these animals and develop intricate and subtle ways of relating to them which is a part of their "way of life". This knowledge goes much beyond that of any commodity.

The reindeer is a sensitive animal, and its behavior is influenced by its inter-actions with humans. The herders know about each reindeer in their herds, their medical problems, if any, and gives names to each reindeer used for milking and riding. So deep are the interactions that Takakura (2010) pre-fers the phenomenon to be termed as "familiarity" rather than as "domes-tication" or "tameness". The herders know about the "personalities" of the reindeer: a particular one may be a bit sly, another may be of a restless type, and yet another may be gentle and cooperative. So, the herders enter into a tussle, a compromise, or a ready collaboration with their animals, also suggesting that they do not have absolute control over the latter. Such a rela-tionship has been termed as "symbiotic domesticity" (Beach and Stammler 2006, cited in Vitebsky and Alekseyev 2015). The Eveny and many other Siberian indigenous peoples also have sacred reindeer or *kujjai*. Everyone has his or her own *kujjai*, which acts as a double for that person. When that person is in danger, *kujjai* receives the attack on that person's behalf. The riding reindeer communicates with its rider by means of snorts and grunts, advises her/him, and warns her/him of avalanches and other hazards. The reindeer is sacrificed when its rider dies and on other special occasions. In the Eveny society, reindeers are of the same status as humans, and the sacred reindeer is an "autonomous agent" like a human. Despite the drive against religion during the Soviet era, many of these practices and perceptions con-tinue. In a vast, harsh landscape full of various dangers, the reindeer and humans strive to survive and flourish together. The domestication here is "symbiotic domestication" or "symbiotic domesticity", where the animals receive protection from insects and other parasites, medical attention in case of need, and assured food supply, and in return, give materials (milk, meat, fur, etc.) and services to humans. Domestication has been defined as "continuous control of breeding of a particular animal population beyond one generation". Reindeer live both as domesticated and wild populations with overlap of range and possible crossing between domesticated and wild individuals (Takakura 2010). The human-reindeer relationships in Eveny (Eastern Siberia) and Nenets (Western Siberia) shed light on this symbiotic association. In Sakkyryr Eveny, the herders have the most intimate famili-arity with dairy does, the riding reindeer, and the draft reindeer. The rein-deer that are being trained and those raised for meat come the next. The Sakkyryr Eveny people have specific terms for each of these classes of ani-mals, and individual names for the dairy and riding animals. The name of a dairy reindeer is usually hereditary, and the herders recognize the mother-daughter kinship. Most dairy does and riding deer accept salt from the hands of the herders. Some meat reindeer may come near a human camp to escape mosquitoes in summer or to lick human urine in winter to meet their salt requirements. The Nenet herders also recognize the individual reindeer in their herds.

Coming down from the far north, human-nature relationship assumes a somewhat unique position in the Nayaka community of South India, where Nayaka hunter-gatherers consider anyone (even a non-Nayaka) with whom they regularly share food, water, space, etc. as their kin, and refer to them as *nama sonta*. This common feature of many hunter-gatherer societies was termed by Alan Barnard as "a Universal Kinship System" (Barnard 1981). "Nayaka do not individuate but ... dividuate other beings in their environment" (Bird-David 1999). The Merriam-Webster dictionary defines dividual as something divided among or shared by a number. The Nayakas share the local environment with other beings, such as a particular hill or an elephant, etc. – some of which may overcome their differences from the Nayaka and become one of them ("absorbs into one we-ness"). Beings which are absorbed like this are "devaru". The devaru are usually addressed by kinship terms, such as *ette/etta* (grandmother/father) or *dodawa/dodappa* (big mother/father). However, not all hills, all stones, all elephants (or other animals) are devaru. An elephant that passed through a Nayaka village without doing any harm, or another that a villager faced in the forest, and which "looked him straight in the eye", could be recognized as devaru. Thus, an elephant (or any other animal/hill/stone) becoming a devaru depends on "mutual engagement" and resultant "relatedness". Bird-David (1990) suggested that relations between man and environment with special reference to hunter-gatherer societies need to be understood because these people obtain their food and other resources from wild sources.

5.7 Man-Nature Dualism versus Nature as Kin

Nayaka views of the environment reveal that they are based around metaphors such as "forest is as parent". Such metaphors reflect the "culturally constituted relationship" between two entities: one the *schema* and the other the *object*. In fact, this represents an "economic model" projected "from the domain of the schema to the domain of the object". Metaphors are not only used for "seeing" the world but also provide broad guidelines for carrying out even simple and mundane everyday activities (Lakoff and Johnson 1980/2003). The forest is akin to the big father and big mother (*dod appal dos awa*) to the Nayaka, who are like their sons and daughters. Similarly, the bear is addressed as grandfather by Siberian Yukaghirs (Willerslev 2013). The Mbuti Pygmies of Congo, and the Negrito Batak of Sumatra, also view themselves as living within the forest which is viewed as their mother or father, and which gives them "food, shelter and clothing", without any obligation to give back. On the other hand, the Bette and Mullu Kurumba, who are the cultivator neighbors of Nayaka, view the forest as their ancestor and are bound to them by the bonds of reciprocity. In general, the cultivator and cultivator-hunter groups in Africa visualize the forest as giving food in reciprocation for appropriate conduct. If the descendants propitiate the ancestor, they are blessed with success in cultivation or hunting; otherwise, they are punished. In contrast, in the world of the hunter-gatherers, the forest never punishes by withdrawing food but may punish in other ways as is done by the parents. Because the forest is parent to all of them, all Nayakas are *sonta* (kin) (Bird-David 1990). Nevertheless, even though the cultivator or cultivator-hunter groups think of the forest in a way different from the hunter-gatherers, the kinship view of the forest is not lost with them. These examples of kinship with nature suggest that the overarching influence of western liberal thought which still rests on the substratum of Cartesian "man-nature dualism" is the main reason for ignoring and downplaying the tradition of ascribing intrinsic value in nonhumans or recognizing the presence of "soul" in nature. The Brazilian anthropologist Eduardo Viveiros de Castro is critical of the western liberal bias, and has called for "permanent decolonization", which means that thinking has to come out of the confining boundary of the Cartesian maxim "I think, therefore I am". He further said that western thinking has been averse to accepting the thoughts of "illiterate people", whose knowledge system comprises oral traditions and not written texts (Skafish 2016). This belief remains rooted in many western philosophers such as Richard Rorty, who said that "We western liberal intellectuals should accept the fact that we have to start from where we are, and that this means there are lots of views we simply cannot take seriously" (Rorty 1991, 29). Contrary to this view, Bronislaw K. Malinowski believed that the objective of anthropology was "To grasp the native's point of view, his relation to life, to realize *his* vision of *his* world" (Malinowski 1922, 25, as cited in Willerslev 2013). The concept of "decolonization of thought" has to be viewed from this perspective, and anthropologists should take indigenous animism more seriously, just as it is taken seriously by the indigenous people themselves (Viveiros de Castro and Walford 2011). Anthropology is essential for perceiving the cosmos in a way appropriate for facing the "ecological crisis". This also stresses the need for "de-Hellenization … and decolonization of thought", which needs to redefine "we" for it to become much more inclusive (Skafish 2016).

5.8 Indigenous Worldviews: No Dualism in Nature

Some examples from tribal or other indigenous societies of India and other parts of the world may help illustrate the nature of their perceptions of the nonhuman world and nature as a whole. India is home to 705 indigenous ethnic groups distributed over 30 states and union territories which are officially recognized as tribes (Mamo 2022). These numerous tribes of India, often regarded as the pre-Aryan or non-Aryan original inhabitants of this land, also have a large repertoire of knowledge on the extrinsic values present in plants and animals – as sources of food and medicine, and often as companions. Almost every tribe has a pharmacopeia which contains their knowledge on the medicinal properties of numerous species of plants and several animals. However, their perceptions and interactions with the natural world go much beyond mere appreciation and utilization of the utility/ extrinsic values.

The cognitive abilities of humans were improved during the Pleistocene as a result of living in groups (Richerson and Boyd 2001; Henrich and McElreath 2002). This in turn enabled them to cooperate with distantly related people living in the same group, and to develop emotional attachments with symbolically marked groups (Richerson *et al.*, 2003). This is also likely to have led to a feeling of kinship and emotional attachment with nature. Many tribes and other ethnic groups that live in close proximity to nature have an apologetic attitude during the extraction of resources from plants and the killing of animals for food. Many groups have taboos on harvesting certain plants and killing certain animals. This could either be a total prohibition or could be effective in mating season or during certain life stages such as during pregnancy or infancy. Thus, a continuity between the human and nonhuman world as well as the living and spirit world developed in many indigenous societies. Here, both totemism and animism are operative, the former used to identify a clan or an individual, and the latter to enable the development of a relation between humans and nature (see Sections 5.3 and 5.4). The indigenous communities do not damage or kill the totem plant or animal. They often have various rites and rituals to propitiate the totem organisms to whom they trace their ancestry. Saraswati (1993) showed the evidence of such non-dualistic pattern of behavior, which is also revealed by their myths and folklores. The Merriam-Webster dictionary defines a myth as "a usually traditional story of ostensibly historical events that serves to unfold part of the world view of a people or explain a practice, belief, or natural phenomenon". A myth is also defined as "a symbolic narrative, usually of unknown origin and at least partly traditional, that ostensibly relates actual events and that is especially associated with religious belief" (Bolle et al. 2020). Saraswati (1993) defined myth "as a body of revelatory knowledge of the unseen reality that flows eternally in time and space". He further elaborated by saying that it "is not a moral code; it is a treatise on the divine cosmology. For a traditional man, it is the wisdom of the ancients that remain all-time new" (Saraswati 1993). In many of these myths, there is no distinction between God, humans, plants, animals, and even inanimate natural entities like sky, river, clouds, and others. In several stories, nonhuman creatures take active part in the creation and construction of the world. Humans can also have their origin from plants, animals, and inanimate objects, and enjoy kinship with these entities. Possessing and imparting knowledge is not the monopoly of humans, either. Very often, it is the nonhumans which possess and impart knowledge to humans (Saraswati 1993). Here some myths and folklores of India as narrated by Verrier Elwin – the doyen of Indian anthropology – (Elwin 1949, 1954, 1958) and Saraswati (1993) are presented to illustrate this pattern.

5.8.1 Some Myths, Folklores, and Folk Practices from India

The creation myth of Gallongs – a tribe mainly living in the West and Lower Siang districts of Arunachal Pradesh in North East India – traces the emergence of Earth from the primeval ocean by the works of prawn and crab, the former collecting a pile of dirt and the latter digging a hole to drain the water. The Hill Miris of Upper Subansiri and Kamle districts believe that when worms burrowed through a primordial tree, a large volume of dust dropped down to form the world. Finally, the tree fell and its bark formed the skin of the Earth, its trunk the rocks, and its branches the hills. In a Minyong (a subgroup of the Adi tribe living in the Siang Valley districts) myth, a *mithun* (*Bos frontalis*, also called the "cattle of the mountain") dug a pit through which the water drained and the Earth appeared. In a Taraon Mishmi (inhabitants of hills and foothills between the Dibang, Digaru, and Lohit rivers) folktale, white ants (termites) climbed up a pillar carrying soil in their mouths to create the Earth. In an Apatani creation myth, the Earth – *kujum chantu* – was herself a giant woman. When she died voluntarily, her head turned into snowcapped mountains, her bones into smaller hills, and her fatty buttocks became the Assam plains with its fertile soil (Elwin 1958). A number of myths also trace the nonhuman ancestry and kinship of man, such as from frogs (Dhammai Miji myth), or from a flower (Khampti myth). The great primeval spirits had three children, one a human, another a rock, and yet another a gourd. When the rock child broke open the gourd, the first humans emerged (Singpho tale). In another story (Taraon Mishmi), the first men and women came out of the tusk of an elephant (Elwin 1958). There are many stories of the marriage and sexual union of humans with gods, spirits, real animals (and not humans in animal disguise) like snakes, monkeys, and tigers, and even leaves, trees, and fire. In these myths, humans are not even unique in the possession of knowledge, which more than often came to them from animals. For instance, the birds (Hill Miri myth) or flowers and bees (Bugun myth) taught the first man and woman the art of reproduction; the spiders taught a girl how to weave (Singpho); and the rat the technique of cultivation (Saraswati 1993). In another story (Idu Mishmi), the sparrow taught cultivation to man (Elwin 1958).

Numerous folklores from the other parts of Northeastern India reveal the deep respect of the indigenous communities for the natural world. A Rengma Naga folk tale (Elwin 1958) narrates that at the beginning of time, all living creatures including plants could speak and understand each other.

Whenever humans wanted to kill an animal or cut down a tree, the latter appealed for mercy, and humans did not have the heart to kill them. In order to enable the humans to obtain their food and other necessities, God robbed all the creatures except the humans and dogs of their power of speech. Now the humans could hunt freely and their dogs, being able to speak, could tell them the exact location of the prey. When this resulted in rampant killing and near-extinction of other beings, God took away the power of speech from the dogs, and hunting again became more sustainable. In this story, we discern an attempt to establish an ethical justification for hunting and cutting down trees to obtain food and other necessities, though unchecked exploitation is not given a divine and, therefore, societal sanction. Hence, the dog is robbed of his speech to maintain the "balance of nature". Similarly, in an Idu Mishmi folk tale (Elwin 1958), the sparrow enjoys the right to take as much paddy as it wants from the fields, as it is believed that the sparrow taught humans the art of cultivation. It seems plausible that as the sparrow often came in large flocks to eat paddy, there was a possibility of people killing sparrows indiscriminately, treating them as pests. This oral tradition, therefore, served to provide some amount of protection to this bird by projecting it as a benefactor of the Idu Mishmis. A similar belief is nurtured by some nomads of Assam, who never kill the sparrow as it is considered the "king of birds". Compassion for nonhuman creatures runs deep in many a folk tale or song. A very popular folk song from Goalpara, Assam, gives a heart-rending recital of a male egret or heron caught in a trap set with small fish to lure the former. Similar songs exist in other areas of Northeast India and perhaps serve the purpose of dissuading people from hunting and killing by appealing to their finer senses (Gupta and Guha 2002).

Compassion has made its presence in many societal practices. For example, there is a widespread taboo in India on hunting during the mating season of animals. Hunters belonging to several communities in Cachar, Assam, do not kill deer during March–May, when pregnant females are present in the herd. Most of these hunters also observe taboo on killing the leader of a deer herd or a sounder of wild boar, as it is believed to be bestowed with supernatural powers, and hence killing it is considered a sin. Again, although many people eat herons and egrets, hunting is banned during the nesting season, and their nests, which are common sights on the bamboo groves of most villages, are never disturbed. The birds are treated as the "guests" of the village. The killing of certain animals is taboo among certain groups. For instance, several ethnic groups in Cachar, Assam, who practice hunting, do not kill the crow, the owl, the vulture, the elephant, and certain snakes, while a group of Muslim trappers do not trap or kill the parrot, the owl, the monkey, and the jackal (Gupta and Guha 2002). Members of the Ramo tribe in Arunachal Pradesh do not eat or kill tigers, because they consider it as their brother, while the Tagins (another tribe in the same state) do not share this belief (Dhasmana 1979). All these myths reveal to us that it is common practice for the tribes and many other indigenous communities to recognize intrinsic values in nonhuman entities of nature.

Similar tales also abound among the tribes of central India. For instance, Elwin (1949) recounts a tale of the Baiga tribe of Madhya Pradesh, which says that the fish donated its eyes to allow God to make pots, and the snake allowed humans to use it as a churn. God then put water in the pot and churned it to make 32 kinds of Earth. The metals also came from the primordial fish: gold from roe, iron from liver, silver from scales, and copper from skin. In another tale, the squirrel dug up the Earth to bring salt and give it to humans. The creation myth of the Santal tribe of eastern and central India narrates that when crab, crocodile, alligator, eel, prawn, earthworm, and tortoise were assigned by God to create land in a world full of water, only the earthworm succeeded in bringing up soil to lay it on the back of the tortoise to create land. Not only was land created by animals, the first humans – a male and a female – were hatched from two bird eggs. Thus, animals to whom the Santals trace their ancestry assume a place of great importance in their cultural life. They consider the totem animals as members of their respective clans, and do not kill or harm them (Raj 2019). Because the two bird progenitors of humans built their nest on the *Karam* tree (*Mitragyna parvifolia*), it is considered sacred and is worshiped by the Santals (Tank 2019).

5.8.2 Myths and Folklores from Other Countries

Brazil (2000) recounted the Australian aboriginal belief that the land is mother, and she "nurtures and nourishes her tenants". It is a reciprocal relationship, land ownership vested in an individual in a hereditary manner also calls for responsibilities of the beneficiary to the land. This "caretaking responsibility" includes duties to maintain the sacred sites and perform traditional rites. He termed it "a mythopoetic relationship" with the land where aboriginal thinking, past and present, reality, and dreamtime are connected like two adjacent rooms. Brazil described the response of an aboriginal person to his land by talking and singing to the rock, and recounting the stories associated with them in a way as if he is "living them, and actually seeing them" in front of his eyes. They recall and represent in art various mythical animals such as the "rainbow serpent" which attracts the rain clouds that are sometimes carried by the *Kirrkarlanjii* (brown falcon) and moves water from one water hole to another in the water-starved desert country (Wroth 2020; Robinson 2022). Here, in these tales, animals and plants are revered for their pivotal role in forming the Earth, and in transporting water. They are in no way subordinate or subservient to humans. Hoefle (2009) suggested that in Amazonia of Brazil, an enchanted worldview regards humans, nonhuman organisms, and inanimate environmental entities such as land, water, or rocks and mountains as living. Humans retain their connections with dead ancestors. On the other hand, disenchantment engenders human-nature dualism, and is the outcome of the spread of reductionist, materialistic science. The belief in the existence of enchanted forest spirits like *Curupira, Mapinguari, Juma, Martin, Saci Pererê*, along with enchanted giant snakes, pink dolphins, and some other animals, is likely to prevent or reduce unchecked over-fishing and over-hunting. The belief that land and water are alive can limit economic over-exploitation. There is a Malayan shamanistic belief that

> *Semangat* (Spirit of Life) is not limited to animals. It permeates the universe, dwelling in humans, beasts, plants and rocks. The universe teems with life: the life of a fire is swift and soon burns out, a rock's life is slow, long and dreamlike.
>
> *(Laderman 1991, 41: cited in Janowski 2020)*

The concept of a vital force or cosmic power driving the cosmos (*Kasektèn* or *sekti*: from the Sanskrit word *šakti*, which refers to the energy or power or force emanating from a deity or more specifically, a female deity embodying and personifying power or energy) is found in many cultures of south and southeast Asia (Janowski 2020).

5.9 Conclusions

The folk beliefs, customs, rituals, and practices, along with the myths and folklores of indigenous societies worldwide, and the hunter-gatherer societies living at the present time (and presumably in the past as well), provide evidence that the worldviews of these societies are based on non-dualism between humans and nature, the latter including its plants and animals, and its inanimate components such as the sky, waters, rocks, and other natural elements. However, the development of objective scientific knowledge in the western societies, which then spread worldwide through their colonies, led to the acceptance of "human-nature dualism" as the dominant worldview by large sections of people in all countries. This expansion also caused an erosion of traditional beliefs and knowledge in indigenous societies, many of which also started accepting the concept of dualism and the consequent large-scale and mostly unsustainable extraction of natural entities, which lost their traditional elevated status of sacredness and began to be viewed as mere resources. However, the prevalence of non-dualistic worldviews in indigenous societies provide support for the contention that the ecocentric ideal of recognition of intrinsic value in nonhuman nature has not emerged only through the modern environmental philosophies, but have a deep-seated presence in traditional societies of the past.

6

NATURE IN RELIGIONS

Venerating Nature

6.1 Introduction

The tradition of humans venerating nature goes deep into the past. Animism and totemism are still prevalent in many indigenous societies all over the world, though these beliefs and practices have become increasingly obsolete and abandoned by modern humans. Nevertheless, traditional conservation of nature characterized by the maintenance of sacred groves, observance of taboos, and other ritualistic practices are still to be found in many indigenous communities. The oral literature, including myths and folklores, and the traditional knowledge systems also reflect strategies oriented towards conservation and sustainable utilization of resources. These practices, which have ecocentric roots in terms of recognition of intrinsic value in nature, appear to have largely succeeded in maintaining biodiversity-rich and undisturbed ecosystems in the areas inhabited and controlled by these communities. A brief overview of the presence of reverence for nature and its veneration in many religions is given in this chapter.

6.2 Veneration of Nature in Indigenous Societies

Sir Edward Burnett Tylor (1832–1917) is one of the first modern researchers to report the practice of water worship among the indigenous communities of Australia, North America, Africa, and Asia. Giving offerings to springs and wells was a common practice in several European countries like Estonia, Bohemia, England, Scotland, Ireland, and France, among others. Despite the strictures of Christianity against such rites and rituals, water worship had persisted "with a varnish of Christianity and sometimes the substitution

DOI: 10.4324/9781003481058-6

of a saint's name" (Tylor 1871, 296–300). Tylor also spoke about a "stage of thought" where a tree was regarded as a "conscious personal being" either possessing a "proper life and soul" or some kind of a fetish or spirit. Such beliefs in trees having a conscious life are also prevalent among many indigenous communities of Asia in the Malaya Peninsula, Borneo, Sumatra, Tonga Islands, India, Myanmar (Burma), Thailand (often under Buddhist influence), as well as in North America, Africa, and Europe. He also noted that tree worship could be in the form of worship of a single tree or a group of trees in a sacred grove. As in the case of water worship, tree worship persists in different forms despite "a crusade against the holy trees and groves" brought about by Christianity (Tylor 1871, 301–305, 313–314). Tylor interpreted these practices as the "illustrations of man's primitive animistic theory of nature" (Tylor 1871, 301).

Another early anthropologist Sir James George Frazer (1854–1941) is best known for his multi-volume compilation *The Golden Bough* (1890–1915). Notwithstanding all the critiques and reevaluations of Frazer's works, *The Golden Bough* remains the most voluminous and extensive compendium of animistic worship of nature. Frazer's "study of magic and religion" has been subject to intense criticisms, debates, evaluations, and reevaluations. In this book, we confine ourselves to the chapter on Tree Worship to gauze its prevalence in ancient communities. Frazer described the practice of oak worship by Celtic druids, and the sacred groves of Germany, where stiff punishments were meted out to anybody hurting the trees in the grove, such as by peeling their barks. Sacred groves also existed in many other parts of Europe such as Upsala in Sweden, Lithuania, Finland, parts of Russia such as the Volga, and in the Slav territories. Inflicting any damage on the trees in the grove – even just breaking a twig – was considered a sin. Such restrictions within sacred groves are also commonly observed in India, and will be discussed in detail later in this chapter. Frazer described tree worship in Greece, Italy, among the Amerindian tribes, different communities in Africa, Thailand (where it is greatly influenced by Buddhism), and other Asian countries. He also observed that some types of trees, or some particular trees, were given more importance than others. For example, it was believed in Dalmatia that particular oaks or beeches are endowed with "souls". Just as the oak enjoyed a special status in Europe, the silk cotton tree (*Bombax* spp.) is revered in West Africa, and the coconut in East Africa. Certain lime, ash, or elm trees were regarded as "guardian trees" in Sweden. Many other species got similar reverence in other parts of the world. The trees were often coaxed and cajoled, or even threatened with dire consequences, if they failed to bear fruit (Frazer 1890, 105–114). Evidences for tree worship are there in *Old Testament*. In Palestine, trees near springs were believed to have divine quality, and the land was considered sacred. Cloth hangings and other offerings were made in such places. However, tree worship was criticized by the great prophets and considered as "something evil and offensive to Jehovah". Subsequently, the practice was gradually abandoned.

Frazer's accounts have been vindicated by publications appearing more than a hundred years after the publication of *The Golden Bough*. Carole Cusack writes that trees and water sources were revered by the Celtic and German people in ancient and medieval times. These practices were condemned as "pagan superstition" by Christianity and many Christian saints felled sacred trees to uproot these beliefs (Cusack 1998). Trees were either venerated in sacred groves, mostly comprising oaks; or through the worship of individual trees dedicated to gods. Trees were mostly associated with male gods, and water sources with female divinities. Sacred groves and trees played significant and intricate roles in the pre-Christian Celtic, Gallic, German, Baltic, and Slavic societies. The remnants of these beliefs can be traced till the present times, and this calls for the "perhaps unconscious need for a more complete relationship between forest and contemporary developed society" (Ritter and Dauksta 2006). The Aquitanian people – who are considered as linguistic ancestors to the present-day Basque-speaking people – lived in the area bordering France and Spain during the Roman times. They were nature-worshippers having veneration for mountains, trees, rivers, caves, and certain animals (Eurasiatic.eu 2017). In Estonia, sacred sites such as forests and springs located within forests were kept intact by the people during the Soviet rule as a part of their belief in a form of forest worship called *Maausk*, which has pre-Christian roots. However, these sites are increasingly being threatened by industrial logging, demand for paper, and a flourishing market for wood pellets as a source of biomass energy. An estimated eighty such sites have already been cleared, with another twelve hundred facing various degrees of threat of extinction. Many of these sites are old-growth forests. The destruction of sacred forest sites perhaps reflects the gradual overtaking of intrinsic value by extrinsic priorities, and commercial interests eroding beliefs and values that had long withstood the influence of Christianity (Elbein 2020).

Moving from Europe to Asia, the Bulang people in Yunnan province, southwestern China use traditional knowledge to protect forests and wildlife. However, the traditional system of conservation is facing threats from an increase in bird-watching tourism. The Bulang people especially venerate the banyan tree and consider it sacred because of its abundant leaves and branches, a spreading crown, high survival ability, and well-developed roots with some roots protruding out on the ground and often having unique shapes. They believe that the high vitality of the banyan tree owes it to its divinity, and if the trees flourish, people will prosper. Conversely, if the trees decline, the people will experience hard times. They are especially reverent of old banyan trees, which are regarded as "treasure of the town" and as "God". Because of these beliefs, the Bulangs have been protecting banyan trees since the ancient period. Besides trees like banyan, Bulang people offer protection to birds, especially the Derbyan Parakeet, which is believed to be "a symbol of mysterious power and wisdom", because they can predict disasters like earthquakes by virtue of their divinity. These parrots build their nests in the holes in old banyan trees, which are often located close to the houses of the villagers (Su et al. 2020).

It is estimated that there are ~13,270 sacred forest sites including sacred groves (SGs) in India, some of which are located in the catchment areas of rivers and streams, while some others are at the sources of rivers or on hill slopes (Rao 1996). However, Malhotra et al. (2001) estimated that their numbers could be over 50,000. These authors found that sacred groves in India – which represent an ancient institution – could be "multi-species, multi-tier primary forest" or could even comprise relatively small assemblages of trees. The groves are usually protected through taboos and sanctions by the local communities and "dedicated to their ancestral spirits or deities". The Scientific Committee on Problems of Environment (SCOPE), Government of India, listed the number and other associated features of the sacred groves in the different states of India. The groves have fruit, timber, and other economically important trees as well as species of religious and ritualistic importance. A large majority of SGs are associated with temples, Buddhist monasteries (*gumphas*), *guruduwaras*, pilgrimage sites, graveyards, memorial sites, churches, etc. Both tribal and non-tribal communities protect and maintain these sacred groves. Presiding deities of sacred groves include *Shiva*, a Vaishnavite (associated with *Vishnu*) deity, several goddesses such as *Amman in* Tamil Nadu, and goddesses of fertility and good health. Many groves are also associated with ancestor worship, and worship of trees and other plants, snakes, and other animals. Trees, other plants, and even animals in sacred groves are accorded intrinsic value by the people and offered total protection. The removal of dead wood is prohibited in many groves and it is allowed to decompose there (Syngai 1999; Malhotra et al. 2001; Singh et al. 2003; Swamy et al. 2003). Peoples et al. (2016) have conducted a study among the present-day hunter-gatherers, the results of which indicate that animism is the oldest trait of religion present in the most recent ancestors of the present-day hunter-gatherers. Of the 33 hunter-gatherer societies investigated, animistic beliefs are present in all, followed by belief in afterlife and shamanism in 79% of the societies. The latter religious trait is less common among the African hunter-gatherers. The ubiquitous presence of animism suggests that Tylor (1871) was not so wrong when he said "animism is fundamental to religion".

6.2.1 Intrinsic versus Instrumental Values in Sacred Groves

It is often suggested that the maintenance of sacred groves by indigenous communities mainly reflects their desire to be prudent in their resource use. They sacrifice some of their immediate benefits to ensure a steady flow of resource. Therefore, through such behaviors, these communities primarily recognize the "bequest value" or use value of the elements of biodiversity and nature as a whole (Gadgil and Guha 1992; Gadgil 1995). This implies that these communities assign only instrumental value to nonhumans, with an objective to ensure their long-term availability in times of need. However, besides ascribing instrumental value, indigenous people also worship nature through the presiding deities in sacred groves, and through tree or animal (totemic or otherwise) worship. Therefore, they cannot be said to assign only extrinsic or instrumental value to the entities they revere, and consequently, their relationship with nature is beyond mere expectations of some materialistic benefits. They had possibly developed a kinship with nature either together with their recognition of instrumental value in nature through an assessment of the resource potential of their habitats, or even earlier. Kellert (1996) suggested that an altruistic attitude towards nonhumans in many indigenous cultures is often motivated by "sentiments of affinity" and not calculated empiricism. Some tribes of West Bengal, India, have been observed to hold as sacred a tree (*Adina cordifolia*) and a shrub (*Euphorbia neriifolia*), which have no known direct use values (Deb and Malhotra 2001). In the *Bathou* religion of the *Bodos* of Assam, the *sijou* plant (a species of *Euphorbia*) is worshipped as the symbol of the Supreme God (Talukdar and Gupta 2018). The sparrow, the jackal, the tiger, various species of songbirds, and snakes – nonhumans that apparently have no instrumental value to these communities – are held sacred in many indigenous cultures of Assam, Manipur, and Tripura in India (Gupta and Guha 2002). A profound sense of "biophilia" could be at the back of such sentiments and beliefs (Fromm 1973; Wilson 1988). The Meiteis of Manipur and Assam, North East India, even go beyond biophilia to "ecophilia" or "cosmophilia" through the practice of "Chingoiron" – the worship of hills – and "Nungoiron" – the worship of rocks (Singh et al. 2003). In belief systems associated with Korean shamanism, deities could reside – besides other places - in trees, ground, rocks, springs, rivers, and the sea (Rhi 1993). An interesting pattern of the recognition of intrinsic as well as instrumental values can be found among the Khasis of Meghalaya, North East India. They maintain sacred groves of several categories. In sacred groves termed *Law Lyngdoh* (forests of priests), *Law Niam* (ritual forests), and *Law Kyntang* (community forests), no exploitation of any entities, including water, soil, rocks, dead wood, and leaf litter, is allowed, and a number of rites, rituals, and religious ceremonies are performed in these community-protected enclosures. Khasis also believe that their dead ancestors reside in these forests till they find their proper resting places (Dutta 1986; Syngai 1999). The Khasis ascribe intrinsic value to all living and non-living entities inside these groves, and do not utilize these groves for any material benefit. However, they also maintain other categories of protected forests such as *Law Adong* (restricted forest), *Law Risumar* (nurtured forest), and *Law Shnong* (village forest). Limited extraction of materials from these forests is allowed with proper approval of the community. Thus, both intrinsic and instrumental values in nature are recognized by the Khasis.

These examples show that a kinship with nature could have originated in the minds of the people before the recognition of instrumental value, which came after undertaking resource estimation and experiencing resource shortage. However, in some cases, indigenous communities might have initially started protecting a plant or animal because of its instrumental value, which subsequently transcended to a recognition of intrinsic value. Table 6.1 lists some plants which are not only worshipped or accorded magical-religious significance by the Meiteis of North East India, but are also prized for their instrumental value. Today, their "divine" associations are the primary reasons for worshipping or revering these plants, while their instrumental values are of secondary importance (Singh *et al.* 2003).

TABLE 6.1 Plants accorded intrinsic value by the Meiteis of Manipur, North East India, along with their instrumental values (adopted from Singh et al. 2003)

Plant botanical name/ common name	Type of intrinsic value accorded by Meiteis	Nature of instrumental value
Toona ciliata/Toon tree	Magical-religious	Timber tree
Dactyloctenum aegypticum/crowfoot grass	Magical-religious	Fodder grass
Cynodon dactylon/dūrvā or Bermuda grass	Magical-religious	Fodder grass; medicinal
Rhus chinensis/Chinese sumac or nutgall tree	Worshipped	Medicinal
Xylosoma longifolia/Dandal	Magical-religious	Medicinal
Oroxylum indicum/Indian trumpet tree	Magical-religious	Medicinal
Blumea balsamifera/Ngai camphor	Magical-religious	Medicinal; condiment
Isodon ternifolius/three-leaf isodon	Magical-religious	Medicinal
Mangifera indica/mango tree	Worshipped	Fruits; medicinal
Aegle marmelos/bael or wood apple	Worshipped	Fruits; medicinal
Ocimum sanctum/tulsi or holy basil	Magical-religious and worshipped	Medicinal
Terminalia arjuna/Arjun tree	Worshipped	Medicinal
Adhatoda vasica/Malabar nut or vasaka	Not harvested on Sunday, as it is believed to be its birthday	Medicinal

Thus, the extrinsic value of nonhumans is not ignored in indigenous and some other societies, though that does not prevent intrinsic value to be simultaneously recognized in the same plant or animal or in inanimate entities like springs, rivers, or forests. In other words, there is no airtight compartmentalization of extrinsic and intrinsic values in a given entity. For example, though the *tulsi* plant (holy basil: *Ocimum tenuiflorum*) is revered and worshipped throughout India, it is also greatly valued as a medicinal plant.

6.3 Intrinsic Value in Nature in Major Religions

6.3.1 Hinduism

Instances of the recognition of intrinsic value in all beings can be found in the *Upanishads* (ancient Indian texts written in Sanskrit – an old Indo-European language), which contain the realizations of Indian seers (*rshis*). There are ten principal *Upanishads* (or according to some 13) believed to have been written between the mid-5th century to the 2nd century BCE. A verse of the *Chhandogya Upanishad* proclaims that "The essence of (all) these beings is the Earth; The essence of the Earth is water; The essence of water is plants; The essence of plants is a person; The essence of a person is speech; and so on" (Swami Nikhilananda 1959: I.i.2, 113). This hymn or verse may be interpreted in the sense that the Earth provides support to all beings; water pervades the Earth and sustains the growth of plants, which in turn supports humans, who are characterized by their speech. Though we might say that in this verse, water and plants are viewed as possessing extrinsic value by virtue of their overriding importance for humans, the close links among these elements show their interdependence with the ultimate support coming from the Earth. This verse also accords a unique status to humans because of their ability to speak. In contrast to this, the *Aitareya Upanishad* proclaims that the self, which is essentially the consciousness, the vital force, and the conscious soul, exists in the Gods, in the five elements of Earth, air, space, water, and fire, in the big and small creatures,

> that are born of eggs, of wombs, of moisture, of the earth, viz. horses, cattle, men, elephants, and all the creatures that there are which move or fly and those which do not move. All these have Consciousness as the giver of their reality; all these are impelled by Consciousness; the universe has Consciousness as its eye, and Consciousness as its end. Consciousness is Brahman.
>
> *(Swami Gambhirananda 1957b: III.i.3, 70)*

This verse perceives the presence of consciousness within the inanimate natural elements as well as all kinds of living things that populate the Earth.

In other words, intrinsic value is recognized in all entities irrespective of their animate and inanimate nature. Finally, the self is described in *Ishā Upaniṣad* as: "That moves, that does not move; That is far off, that is very near; That is inside all, and that is outside all". In the next verse, it follows that "He who sees all beings in the very Self, and the Self in all beings, feels no hatred by virtue of that (realization)" (Swami Gambhirananda 1957a: 5 and 6, 11–12). In these two verses from the *Ishā Upaniṣad*, we can see an assertion of realization of the unity among all entities – human and nonhuman alike – and the upwelling of love and compassion for all beings. Thus, we detect a progression from extrinsic to intrinsic values. It is believed that the *Chhandogya Upanishad* was one of the first *Upanishads* to be written, followed by the *Aitareya* and then the *Ishā* . Therefore, it may be possible to theorize that the recognition of extrinsic/use value in nature came first, followed by a transcendence to a deeper realization of the inner unity that exists within and between the living and the non-living world, with humans as one of the elements of this assemblage. If we take the *Vedas*, which are ancient texts that were written during ~1500 to 800 BCE, Hymn XC of the *Rig Veda* – the earliest of the *Vedas* – has a prayer: "So may the plants be sweet for us, ... May the tall tree be full of sweets for us" (Griffith 1889: Hymn XC, 48). *Rig Veda* (*Book the Fifth*), Hymn XCVII is called the "Praise of Herbs". It addresses the herbs as mothers and wishes them to "have a hundred homes" and a "thousand growth". It pleads with the herbs to have "thousand powers" and cure the patients of disease. It compares the person who possesses a store of herbs to a king, and praises the herbs rich in "soma" (drink), in "nourishments" and "strengthening power", in "healing virtues" to give a person "wealth" and "save (his) vital breath". It also names them as "restorers". They bring back the "vanished strength" and the "spirit of disease departs", and the hymn showers more praises and prays that the herbs bestow good health and prosperity upon humans and domestic animals (Griffith 1889: Hymn XCVII, 474–475). In the fourth and the last of the *Vedas* – the *Atharva Veda* – plants are invoked to become auspicious for curing hereditary diseases. There is mention of an aquatic plant *Blyxa octandra*, the sharp spikes of which are believed to drive away evil, and the magical properties of amulets made from the *sraktya* tree (*Clerodendrum phlomoides*) (Bloomfield 1897). Thus, in these hymns, the material importance of plants in supporting a good life is portrayed, or in other words, there is cognizance of their extrinsic value. However, the *Yajur Veda* – the second *Veda* – sings praise to the plants in several hymns (Keith 1914). For example, Hymn i.3.9 (Keith 1914, 646) hails the plants along with water and Earth; hymn i.8.13 (Keith 1914, 714) embraces the cosmos and nature when it hails to the Earth, the atmosphere, the sky, the sun, the moon, and the Nakshatras (stars). It also hails to the waters, plants, trees, and the moving, swimming, and creeping creatures. Interestingly, while hailing the plants along with their roots, twigs, flowers, fruits, etc., hymn vii.3.19 (Keith 1914, 1146) also hails to fruits (or plants?) that are used and not used, and to those that have fallen off and those that are lying on the ground. In the next verse (vii.3.20), there is a similar prayer, which also includes those that are left over, left out, deprived of or with leaves, and those deprived altogether. These hymns recognize value in the rather insignificant entities such as even those fruits or plants, which are useless, fallen, or barren, or the creeping creatures, which are otherwise ignored or at times abhorred. It is very natural to praise or appreciate something spectacular or highly useful, because it stems from our fear or apprehension that these may not be available or accessible to us some day. On the other hand, when we appreciate something of no known use or of no great beauty, what is wasted or spoilt, or even abhorred, we are doing that because we want them to be there just for their own sake. This also shows that the extrinsic values of plants were well-recognized because they served as food, and were essential for treating various diseases afflicting people. However, this did not prevent the realization of the intrinsic values of nonhuman creatures such as plants and animals, and even the Earth and the cosmos as a whole. This also shows that recognition of intrinsic value in nonhuman entities also emerged during the writing of the *Vedas*, culminating in their fullest and the most poetic expressions in the *Ishā Upanishad*.

The recognition of intrinsic value of plants is also found in *Manu Samhita* – an ancient Hindu text dating to ~100 CE – a few centuries later than the *Vedas* and *Upanishads* – which says that the plants have a "dormant or latent" consciousness, and can experience pleasure and pain. Other texts and philosophical commentaries dating to ~1000–1500 CE also speak of the presence of an elementary, "stupefied", "comatose", or "unmanifested" consciousness in plants. Gunaratna, a 15th-century scholar, referred to the similarity of humans and plants by virtue of their possession of consciousness (Majumdar 1927). Thus, it seems that a general perception of the presence of inherent or intrinsic values in plants was prevalent in ancient India, and it persisted from the Vedic ages to medieval times (~1500 CE). To say this is not to deny the fact that the extrinsic values of plants (and animals) were also highly appreciated in ancient India, where a large body of knowledge existed on their applications for curing various diseases, besides the knowledge on numerous edible plants. This is testified by the fact that around 7500 plants are known to be used in Indian traditional medicine, of which about 1769 are used in *Ayurveda*, 743 in *Siddha* (system of traditional medicine in South India), 653 in *Unani* (a tradition of Graeco-Arabic medicine, widely practiced in India), and 4671 in folk medicine, mostly practiced by the numerous tribal and other indigenous communities. Besides, about 76 species of animals are used all over India in tribal medicine. There is a lot of overlap in the use of the plants among these different systems, though the same plant may also find different uses in the various systems of traditional medicine (Gupta 2004). Thus, the awareness of the presence of both extrinsic and intrinsic values in plants existed for a long time in Hinduism. This trend has persisted in Hindu religious traditions, and is depicted in the *Vaishnavite* expression of *Trinad api sunichena, tarur iva sahishnuna*, which implies that a person should be lower (or humbler) than the grass, and (more) tolerant than the tree. The religious precept of Srimanta Shankardeva, the Vaishnava saint from Assam (1449–1569), who initiated the *neo-Vaishnavite* movement in the 15th century CE, not only overcomes the caste or religious barrier, but goes beyond the human species to embrace other forms of life (Gupta 2013). Shankardeva said "*Kukkuro candalo gardabhoro atma ram, / Janiya sabako pori Kariba pranam*", which asks everyone to recognize and respect the presence of godly soul in the dog, the donkey, and a person relegated to the lowest strata of the caste hierarchy (Das 1978). Here, ecocentrism can be seen to unite all humans and take the nonhuman world along, giving it an equal status.

6.3.2 Jainism: No Harm to the World

Besides *Hinduism*, India is also home to the religious tradition of *Jainism*, which is a heterodox response to the ritualistic approaches of Hinduism. Mahavira (~599–527 BCE), who is commonly regarded as the founder of this branch of ancient religion, is regarded by the Jains as the 24th and the last *Tirthankara* (enlightened one). It has two major precepts that have implications for the recognition of intrinsic value. These are: *ahimsa parmo dharma* (non-violence/compassion is the supreme virtue/religion); and the other is *parasparopagraho jivanam*, which implies that living beings render service to one another or, in other words, depend on one another. *Jainism* has the concept of *jiva*, which not only denotes living organisms, but can be said to simultaneously represent the soul or *ātmā*, and is characterized by consciousness, bliss, and energy. *Jivas* are classified into two major groups on the basis of mobility: immobile or *sthāvar jivas*, and mobile or *trasa jiva*. The immobile or *sthāvar jivas* are also called *ekendriya* or single-sense *jivas*, which only have the sense of touch. These are further classified into five major types:

i)   *prithwikaya* or earth-bodied, which include clay, sand, metal, coral, etc.;

ii) *apakaya* or water-bodied, including different forms of water such as dew, fog, rain, ice, etc.;

iii) *teukaya* or fire-bodied, including fire, lightning, flame, hot ash, etc.;

iv) *vayukaya* or air-bodied, such as wind, whirlwind, storm, etc.; and

v) *vanaspatikaya* or plant-bodied, which include herbs, shrubs, trees, branches, flowers, leaves, seeds, etc.

Thus, *Jainism* recognizes life in the things that we consider as inanimate, along with the plants.

As opposed to the immobile, *sthāvar jivas*, there are the mobile, *trasa jivas*, which are *bahu indriya* or multi-sensed. These are of the following four categories:

i) *beindriya jiva* or two-sensed beings with sense of touch and taste, such as shells, worms, some insects, microbes in stale food, etc.;

ii) *treindriya jiva* or three-sensed beings with sense of touch, taste, and smell, for example, bugs, lice, moths, insects in stored grains, etc.;

iii) *chaurindriya jiva* or four-sensed beings with sense of touch, taste, smell, and sight, such as scorpions, crickets, flies, spiders, beetles, locusts, etc.; and

iv) *panchendriya jiva* or five-sensed beings having sense of touch, taste, smell, sight, and hearing, for example, fish, birds, cow, lion, human, etc. The nonhuman five-sensed beings are called *tiryancha* to distinguish them from *manushya* or humans (URL 3, URL 4).

Thus, there is recognition of soul – or in other words, intrinsic value – in both inanimate and all animate things, albeit with some kind of gradation or hierarchy among them, with the humans at the apex. However, the humans are not given the right of dominion over the other *jivas*, because one of the two cardinal teachings of *Jainism – ahimsa parmo dharma –* prohibits any act of violence towards any organism, while the other – *parasparopagraho jivanam –* emphasizes the interdependence of all forms of life. The respect for non-human life is evident in the following teaching of Lord Mahavira, where he advised the *Jain* monks and nuns not to view big trees as "fit for palaces, gates, houses, benches ..., boats, buckets, stools, trays, ploughs, machines, wheels, seats, beds, cars, and sheds". Instead, these should be regarded as "noble, high, round, with many branches, beautiful and magnificent" (*Acaranga Sutra* as quoted in Chapple 2001). Thus, Jainism recognizes these trees to be valuable in their own right as beautiful and majestic members of the natural community.

6.3.3 Buddhism: Messages of Universal Love or "Metta"

Prolonged and intricate discussions about the possession of intrinsic value by the natural world have taken place among Buddhist philosophers, only a glimpse of which is given here. La Fleur (1973) pointed out that many Chinese and Japanese Buddhists have made elaborate discussions on whether "plants and trees", and in fact, the whole natural world, could "attain Buddhahood", or in other words, salvation. Chi-t'sang, a Chinese Buddhist thinker, stated that since plants and trees are "essentially like sentient beings", they have the capacity to attain Buddhahood. Chan-Jan (711–782 CE), a Chinese Buddhist thinker suggested that even nonsentient beings could possess the "Buddha-nature". He asserted that there was no essential difference among grasses, trees, and the soil from which they grow, and all could reach *Nirvana*. Hence, even inanimate things (such as soil) are not devoid of "Buddha-nature". This, in fact, is an extension of the Mahayana principle that even non-believers (in Buddha) could attain Buddhahood. Kūkai or Kōbō Daishi (774–835 CE) – the founder of Japan's Shingon School of Buddhism – spoke about "the Buddha-nature of Trees and Rocks". He also said that though the physical eye can only discern the coarse form of plants and trees, the "Buddha-eye" can recognize their "Buddha-nature". This precept is taken one step further when Ryōgen (912–985 CE) – a well-known Japanese Buddhist monk of the Tendai School – reasoned that since grasses and trees have four phases, that is, sprouting out, growing, reproducing, and dying, they could be said to "aspire for the goal, undergo disciplines, reach enlightenment, and enter into extinction". However, it must be said at this juncture that some Buddhist schools of thought refute these concepts of granting "Buddha-nature" to the natural world. Besides these direct references to intrinsic value in the natural world, the Pali word *mettā*, which denotes universal love and loving-kindness among its several other meanings, is put forward in the *Karaṇ īya Mettā Sutta* – which contains the teaching of *Gautama Buddha* – though it might have been put down in the written form only around 5th century CE. The *Sutta* says:

> May all beings be happy and secure, / May all be well-disposed at heart! / Whatever living beings there be, / Without exception, weak or strong. / Long or huge, middle-sized, / Short, minute or bulky. Whether visible or invisible, / And those living afar or near, / The born and those seeking birth, / May all beings be happy! .... Let him cultivate a mind of boundless love, / Towards all, throughout the universe, / In all its height, depth and breadth, / Love that is unobstructed / And free from hatred or enmity.
> *(Buddharakkhita 2009)*

Since the *Sutta* prays for the flourishing and well-being of all beings, and preaches "boundless love" towards the whole universe, recognition of intrinsic or inherent value in plants, animals, and the whole natural world can be said to be implicit in its teaching. The Maurya emperor Ashoka (period of reign: 273/272–236/235 BCE) ran a victorious but bloody campaign in Kalinga (present-day Odisha and northern Andhra Pradesh) in which an estimated 100,000 to 250,000 people were killed during and after the war.

It is said that stricken with remorse at this bloodbath, Ashoka eschewed violence and took to Buddhism, the teachings of which he was instrumental in spreading throughout Asia. We also might detect an overwhelming sense of *mettā* in this ruler, who not only gave up violence against his human subjects, but also drastically reduced the number of animals killed and fish captured in his kingdom through the promulgation of the "Law of Piety". Through his rock and pillar edicts, Ashoka also incorporated these prohibitions and general guidelines for compassionate behavior towards animals into the Maurya State policies. In his pillar edict (PE) 5, he declared many animals such as several species of birds, bats, turtles and tortoises, fish, some species of mammals, and "all four-footed creatures that are neither useful nor edible" as protected. He also ordered that "husks hiding living beings are not to be burnt and forests are not to be burnt either without reason or to kill creatures" (Dhammika 1994, Gupta and Ghosh 2003). Thus, though there was no declaration as to the recognition of intrinsic value in non-human creatures, the practice followed by the Mauryan State under his rule reflected the recognition of such value. Mukherjee (2000) rightly commented that the *dhamma* of Ashoka included "even the animal kingdom within the scope of its all-embracing benevolence".

6.3.4 Shinto: The "Way of Kami"

Of the three major religious philosophies prevalent in Japan, *Shintoism* is the most prominent and perhaps wields the greatest influence. It embodies the traditional religious practices and belief systems of Japan. The origin of *Shinto* could be traced back to 3rd or 2nd century BCE. The word *kami* is often translated as God, though the concept of *kami* and a western god are fundamentally different. The polytheistic faith of *Shinto* recognizes kami in mountains, seas, ancestors, kings and other prominent personalities, animals, and others. Shinto has also not evolved in isolation: Buddhism, Taoism/Daoism, and Confucianism have their influences on and interactions with Shinto. Monotheistic traditions like Christianity as well as modern western ideas have their influence on Shinto as well. It is often said that there are eight million *kami* in *Shinto*. However, it is not a fixed pantheon, because *kami* can exist or be manifested in myriad natural entities. For example, celestial bodies such as the sun, moon, and stars are worshipped as *kami*; forests, lakes, mountains, rivers, stones of unusual or unique shape, etc. are regarded as housing *kami* or are themselves regarded as *kami*. *Kami* is also manifest in many animals such as snakes, bear, fox, crocodile, deer, wolf, tiger, monkeys, crow, etc. (Nobutaka 2000). The oldest Shinto texts including the *Kojiki* and *Nihon Shoki* name Amaterasu, the sun goddess, as supreme amongst the gods or *kami*, with others such as Susanoo, the wind and sea god, etc. (Cartwright 2017). The 18th-century Japanese scholar,

Motoori Norinaga explained that *kami* denotes the deities of heaven and Earth, many of which are also worshipped in shrines. Besides human beings, "birds, beasts, trees, plants, seas, mountains, and so forth", dragons, foxes, tiger, wolves, rocks, stumps of trees, and leaves of plants, particular mountains or seas are also considered as *kami* (Holtom 1940). Though Shinto gods are worshipped in shrines, their existence is not different from natural features – mountains, seas, forests, and others – they are parts of nature. In the words of H Byron Earhart, "nature intrinsically manifests the sacred or is *kami*" (Earhart 1970). In other words, not only does nature harbor *kami*, it is itself *kami*. This is particularly exemplified by the mountains of Japan, which are treated by the Japanese with deep reverence. Mount Fuji, Mount Yatsugatake, Mount Aso, Mount Hakusan, Mount Kurikomayama, Mount Tadagatake, the Miwayama mountains, and many others are examples of sacred mountains representing or housing *kami*. These mountains can be of different height, shape, and origin. They can be stratovolcanoes, which are typical cone-shaped with steep flanks such as Mount Fuji; or other types of volcanoes (e.g., shield volcanoes with more gentle slopes) such as Mount Aso in Kumamoto, which is an active volcano. Snow-capped mountains not of volcanic origin, such as Mounts Hakusan, Kurikomayama, and several others, are also regarded as *kami*. And the third category is called the *Kannabi* mountains, which broadly denotes an abode of divinity. The height of these mountains is not a factor in their sacredness; some of them could be just a few hundred meters tall. Some of the properties that these mountains or hills should ideally possess to be recognized as a *kannabi* include a regular shape, proximity to a human settlement, and dense forest cover. A river flowing in its vicinity is also an important factor. The Miwayama in the Yamato basin in Nara prefecture is a classic example of a *kannabi* mountain. The whole mountain can be said to serve as a shrine. However, it is also to be understood that such sacred sites and the concept of their sacredness also represent a harmonious fusion between Buddhism and Shinto in Japan. In that sense, they are all the more valuable from a pluralistic ecocentric viewpoint (Yano 2008).

Thus, according high intrinsic value to non-human inanimate nature as well as plants and animals is a cardinal precept in Shinto. This is notwithstanding the fact that the realization of intrinsic properties could have transcended from a deep sense of gratitude for the many extrinsic values of nature, on which the ancient Japanese (as well as human societies elsewhere) were dependent for their survival and sustenance. This is reflected in a poem by Motoori Norinaga:

The foods we eat, / The trees and grasses, / Are all vouchsafed / Through the blessings / Of the Sun Goddess. / Without the blessings / Of the kami / Of heaven and earth, / How could we exist / For one day, for one night?
(Hirai 1960)

6.3.5 Dao and Confucian Thoughts

6.3.5.1 Confucianism: It Is "Man that Makes the Dao Great"

Confucian environmental ethics has often been called "anthropocosmic" and not overtly anthropocentric (Wei-Ming 1998). In such a view, humans are integral parts of "heaven, earth and myriad things", albeit with their "intrinsic capacity of the mind to embody the cosmos in its conscience and consciousness" (Wei-Ming 1985; quoted in Wong 2015). Like the other anthropocentric environmental philosophies, Confucianism emphasizes human values, but has the non-anthropocentric characteristics of these values transcending the human realm. The Confucian *Doctrine of the Mean* asserts that "those who are absolutely sincere" can "fully develop the nature of others" and in turn can "fully develop the nature of things", and then they can assist" in transforming and nourishing "heaven and earth" to thereby form a trinity (Chan 1969, quoted in Wong 2015). However, Ruiping Fan (Fan 2005) is of the opinion that the nomenclature "anthropocosmic" instead of anthropocentric to designate the environmental ethics of Confucianism is "too obscure, ambiguous and imprecise" for providing normative guidelines for human attitude towards and action on animals and nature as a whole. Though Confucianism recognizes that only humans have intrinsic value, it transcends the level of the humans and the environment, and is guided "by the cosmic principles disclosed by the Confucian sages" (Fan 2005). Tucker (2001) went further to suggest that Confucianism accords nature inherent value, and hence, is compatible with the basic tenets of Deep Ecology. On the other hand, Nuyen (2011) maintained the position that Confucianism recognizes inherent value in nature, but not intrinsic value. Nuyen distinguished between inherent and intrinsic values in the following way. Something has inherent value if the "basis of the value" is in the entity itself, but the value is imparted by outside agents, such as humans, or in other words, human valuers are essential. On the other hand, when intrinsic value exists in any entity, it is not dependent on human valuation. Inherent value is non-instrumental, because it imparts a value that is not confined to human utility or preference, and yet is dependent on humans for its valuation. Any ethic that recognizes inherent value in nature, but does not accord it intrinsic value, is still anthropocentric, although its anthropocentrism is of a weak variety. Nuyen (2011) referred to the ancient Confucian scholar Xunzi (3rd to 2nd century BCE) who said that humans form a great triad with heaven and Earth, and therefore, human identity is shaped by the nonhuman components, and at the same time humans bring order to heaven and Earth through the moral rules framed by them. Confucius also forbade people to do anything to further their interests or well-being by violating the *Dao*, which implies that the value of *Dao* is inherent, though it is not intrinsic, because Confucius says that "it is not the *Dao* that makes man great but rather man that makes the *Dao* great". As put succinctly by Miller (2003) and also interpreted by Nuyen (2011), *Dao*, which means "way", is synonymous with "nature" or the "flourishing of nature itself". In the words of Lai (2015), the Confucian doctrine of the linkages of humans with heaven and Earth places humanity "in a larger cosmological perspective", thereby elevating its status. All these statements mean that while Confucianism has great respect for nature in recognizing its inherent value, it nevertheless attaches the greatest importance to human appreciation and treatment of nature to guide life along the moral path, and in that sense, does not totally abandon anthropocentrism.

6.3.5.2 Daoism: "Heaven and Earth Are Not Focused on Humaneness"

Besides Confucianism, the teachings of *Daoism* or *Taoism* dating back to ~6th century BCE have wielded considerable influence on the ethical norms accepted and practiced by the Chinese society. *Daodejing*, a well-known *Daoist* classic, has an "anti-anthropocentric tenor", and advances the basics of an environmental ethic that "reaches beyond humans, individuals, or species" (Lai 2003). The ancient *Dao* text of *Laozi/Daodejing* 5 highlighted the equality among entities thus: "Heaven and earth are not centrally focused on humaneness. They regard the ten thousand things as straw dogs. The sage is not centrally focused on humaneness. He regards all people as straw dogs". Here, the importance of being human is underplayed and any elevated status to humans is denied. The verse refers to straw dogs, which are indispensable and treated with deference before they are used as offerings in a sacrifice, but are thrown away and trampled after the sacrifice rituals are over. The straw dog metaphor signifies that neither humans nor nonhumans or any other entity has intrinsic or extrinsic/instrumental value per se, the nature of their value depending on the context of the situation. Instead, *Daoism* stresses the interlinking of humans with the non-human aspects of the environment. *Daoism* presents a perspective where the independence of human or any other being is not recognized, because it will negate the interdependence among all beings. It, therefore, does not brand any entity including humans as having intrinsic or extrinsic value, because they have values which are "neither only instrumental nor only intrinsic", but are dependent on the position of the being within the whole. In other words, it is like the value of the straw dog which is entirely dependent on the context in which it is situated (Lai 2015).

6.3.6 Christianity: St. Francis of Assisi and Laudato Si'

The Christian creation story is distinctly different from the indigenous cosmogonies, some of which are described in Chapter 5. Lynn White Jr. in his commentary on the role of Christianity in conceiving and spreading an

anthropocentric environmental ethic observed that in the Christian creation story, "a loving and all-powerful God" created the nonhuman world
of plants and animals in stages, and this process of creation culminated in
the creation of first Adam, followed by Eve, to dispel his loneliness. Humans
named all plants and animals, which established their dominance over the
latter. All these nonhuman creations are for the pleasure and benefit of
humans. Human supremacy over all creatures is especially ascertained in
Genesis Verse 1: 26, which says

> And God said, Let us make man in our image, after our likeness: and let
> them have dominion over the fish of the sea, and over the fowl of the air,
> and over the cattle, and over all the earth, and over every creeping thing
> that creepeth upon the earth.

White said that "especially in its Western form, Christianity is the most
anthropocentric religion the world has seen" (White 1967). Before
Christianity became the dominant religion in the west, followers of pagan
animism placated the guardian spirits in the forests, rivers, springs, and
mountains before harvesting any resource or tampering with any ecosystem.
This had a moderating influence on the destructive influence of human activities on nature. After the spread of Christianity, people gradually lost their
reverence for nature and felt far less inhibited to extract natural resources
from the natural ecosystems. However, an "alternative view" of Christianity
was provided by St. Francis de Assisi, who denied humans their proprietary
rights over other species and preached equality for all God's creatures. Even
a fierce wolf, which ravaged the Apennines, was "brother wolf" to him. The
Church has declared St. Francis of Assisi as the Patron saint of ecology, serving as a guiding beacon to all environmentalists and ecologists, and all those
who love animals and nature. Pope John Paul II in 1979 cited St. Francis as
"an example of genuine and deep respect for the integrity of creation". In his
"Canticle of the Creatures", St. Francis of Assisi praised the Lord through
all his creatures, holding a special regard for Brother Sun, along with Sisters
Moon and stars, Sister Water, Mother Earth, Brothers Wind and Fire, and
finally Sister Bodily Death, whom none could escape. In his "Letter to the
Faithful", he wrote that every creature on Earth and the depths of sea praises
the Lord (Duddy 2022a, b, c). On June 18, 2015, Pope Francis published his
Encyclical Letter *Laudato Si'* with the subtitle "*On Care for Our Common
Home*". In this he paid a rich tribute to Saint Francis of Assisi and his "beautiful canticle, ... [in which] Saint Francis of Assisi reminds us that our common home is like a sister with which we share our life and a beautiful mother
who opens her arms to embrace us". He further said that "this sister now
cries out to us", and "symptoms of sickness [are] evident in the soil, in the
water, in the air, and in all forms of life". The mother Earth now "groans in

travail". The Pope called Saint Francis of Assisi a personification of "integral ecology", and asserted that the religions can make a "rich contribution ... towards an integral ecology and the full development of humanity". The concept of integral ecology does not view nature "as something separate from ourselves, or as a mere setting in which we live. We are part of nature". In the concluding section of the Encyclical, the Pope laid down "A prayer for the earth", where he prayed to God to "Teach us to discover the worth of each thing, / ... to recognize that we are profoundly united with every creature" (Pope Francis 2015).

6.3.7 Islam

Islam regards nature as a perfect creation of God. Nature has a "well-knit structure" without any imperfection. It "reflects the power, beauty, wisdom and mercy of its creator" (Rahman 1980, as cited in Özdemir 2019). The entire creation including plants, animals, and humans bow down in worship to God. Islam's precepts do not advocate "that the whole of nature and its resources are designed for human benefit alone .... Rather it is a bestowed trust for human beings". Islam, therefore, lays great stress on compassionate treatment of animals, and to take care that they are protected, fed well, and in no way tortured or degraded. If a person plants a tree and it is enjoyed by humans, birds, and other animals, then it is regarded as an act of charity (Gada 2014; Özdemir 2019). The teachings of Islam guide and control human attitude towards consumerism, which causes degradation and destruction of the natural world. These teachings, therefore, serve as a common ground with ecocentric ideals in opposition to anthropocentrism. Islamic concepts of environmental ethics are non-anthropocentric, as pointed out by many scholars (reviewed in Rahman 2022).

6.4 Conclusions

This chapter goes on to show that affinity and reverence for nature run deep in the philosophies of many eastern religions, in Islam, and in certain branches of Christian religious thoughts. Though the monotheistic religions primarily recognize a stewardship role for humans, they also look at humans and nature as the creation of the same God, and therefore, recognizes the sanctity of nature. This also suggests that strains of ecocentric thoughts come naturally to humans and find expression in many streams of religious philosophy spanning over a long time period as well as vast tracts of geographical space.

7
NATURE IN THE REALM OF IMAGINATION

7.1 Introduction

This chapter does not intend to give an overview of the widespread presence of ecocentric thoughts and imageries in visual art, poetry and other genres of literature, and music. Its objective is to suggest that recognition of intrinsic value in nature comes naturally to humans, though not all of us might yet be ready to accept it. However, recognition of the intrinsic value of nature spontaneously comes to artists, musicians, poets, and writers, who depict it in their creative works. This is not only accepted but also appreciated by society because these expressions represent forays into the realm of imagination. Anthropologist Edward Burnett Tylor traced the origins of this intrinsic recognition to the animistic beliefs which transcended from being a "philosophy of life" to a "philosophy of nature" (Tylor 1871, 271). He further said that when people returned to their childhood's boundless imagination, they saw streams, mountains, forests, and other elements of nature as living entities. Such feelings find their fullest expression in poets, and the philosophy of yesteryears that was conceived by the animists became the poetry of today. The biocentric/ecocentric legacy of the ancestors is carried by modern humans, and it is most fluently and spontaneously manifested in the poets. Henry David Thoreau said that it is only the poet for whom "the winds and streams" would speak; "whose words" – "true and fresh and natural" – could stimulate the buds to blossom at the advent of spring even between the dry leaves of a book, because these words are "in sympathy with surrounding nature" (Hartman 2007). Here the word "sympathy" holds the key, since words like sympathy, empathy, or altruism are only used in relation to entities which are regarded as having intrinsic value. This is

DOI: 10.4324/9781003481058-7

a moral-ethical position that the poet (or a writer/songwriter or artist) can reach with ease. When Thoreau says that the poets can make the winds or streams to speak for them, it also suggests that the poet can put words to the unheard voices of winds, streams, unopened buds, wilted trees and suffering animals, and even the entire stripped and tortured Earth. This happens because poets can visualize humans, inanimate nature, and nonhuman living beings alike as entities having intrinsic value – a value that arises from within and not ascribed by someone in their self-interest like a price tag. Just as nature's expressions, aspirations, and woes find a medium of expression in the verses of poets, deprived and oppressed people also find solace and refuge in nature, and the "silent dialogue" between them and nature is the subject of creation of many poets, novelists, and story writers. Especially notable among such oppressed are women, who along with nature itself have often been the subjects of exploitation and oppression by the anthropocentric patriarchy. This struggle and the tortured experiences of both nature and women have found expression in the writings of ecofeminist poets and writers of several genres of literature. Nature appears as a compassionate figure with its own intrinsic properties of compassion to the exploited and oppressed women, and a bond of love and understanding grows between them.

This chapter provides a "green reading" of the works of a few poets and writers known for their identification with nature and their perceptions of a feeling of communion with nature, whereby there is a sharing of thoughts between humans and their living and pulsating environment. Poets and writers writing before the 1960s have been deliberately chosen, because it is only natural that biocentric or ecocentric thoughts would permeate the writings of post-1960 writers because of the spread of awareness about environmental degradation and the need to recognize animal and other nonhuman rights in the post *Silent Spring* period. On the other hand, poets who wrote in the days when human encounters with nature – which more often than not turned into trampling and stripping of the latter – were mostly lauded as "conquests", and the human "conquerors" glorified as pioneers. In such an atmosphere of dominance of anthropocentric thoughts, these poets could build "dialogues" of understanding with nature, thus providing evidence for the belief that biocentric and ecocentric thoughts are also a part of the human heritage, though these might have long been suppressed by anthropocentric images of nature as a mere commodity meant solely for human use and enjoyment. Furthermore, poets and writers have been selected from two distinct languages and countries – English in the United States of America, and Bengali in the Indian state of West Bengal (Jibanananda Das, one of the Bengali poets cited here, was born and grew up in Barisal, Bangladesh, though he got recognition as a poet during the latter part of his life in Kolkata, India). This has been done to emphasize the cross-cultural prevalence of such communions with nature. However, all these examples should be taken as illustrative rather than exhaustive. There is no dearth of other examples in different cultures and languages that show the appreciation of nature and recognition of its intrinsic value in almost all possible forms of creative productions or manifestations.

7.2 Poets and Writers from the West

7.2.1 Ralph Waldo Emerson

Ralph Waldo Emerson (1803–1882) was a transcendentalist who thought that "all natural objects can awaken reverence". His writings often depicted pantheistic and animistic convictions, and he believed that nature wields a spiritualistic influence (Taylor 2012). However, this poet and thinker did not consider himself a philosopher, but "simply as a person who related what he had experienced, … (and) he revealed the world as it was revealed to him". Emerson imagined nature as an entity in which there is the presence of God. He valued the "kindred" view of nature that establishes a close tie between humans and nature. This image of nature is essentially poetic in its meaning. In such an image, not only do "the flowers, the animals, the mountains, reflected the wisdom of his best hour", but also "delighted the simplicity of his childhood". He felt that very few adult humans can perceive the true image of nature. For example, the sun only carries a "superficial" meaning and "illuminates only the eye of the man, but shines into the eye and the heart of the child". "The lover of nature is he … who has retained the spirit of infancy even into the era of manhood". Emerson, therefore, asked humans to take a fresh look at nature, and to understand the capacity of nature to liberate the human spirit. When transformed this way, humans no longer see nature merely as a resource. It is with this vision that the poet sees the "tree" and not a "stick of timber of the wood-cutter". Emerson also said that individuals may own the land in farms or woodlots, but none of them "owns the landscape". It is only the eye of the poet that "can integrate all the parts" of this montage (Emerson 1849, 3).

Despite such perceptions of nature, many researchers identified Emerson as an anthropocentric thinker. Ann Woodlief said that the anthropocentric views of Emerson were reflected in his statement that "All the parts (of nature) incessantly work into each other's hands for the profit of man" (Woodlief 1990). However, Woodlief did not mention what Emerson wrote immediately after this sentence, which is

The wind sows the seed; the sun evaporates the sea; the wind blows the vapor to the field; the ice, on the other side of the planet, condenses rain on this; the rain feeds the plant; the plant feeds the animal; and thus the endless circulations of the divine charity nourish man.

(Emerson 1849, 5)

In these words, Emerson describes the natural cycle that sustains the food chain which serves plants, animals, and humans alike. Thus, when he speaks about the "profit of man", he does not mean commodification of nature for making material profit, but more about the sustenance that humans derive from natural cycles because they are essentially a part of nature. Woodlief also wrote that Emerson saw "nature as a great and holy teacher of the self-reliant man who will look beyond its uses as mere commodity and see it as infused with spirit" (Woodlief 1990). In these words, she contradicts her earlier statement, because an anthropocentric view of nature only values it for its instrumental use. On the other hand, these words of Emerson reflect his ecocentric inclinations. While discussing the anthropocentric and anthropomorphic imageries in the nature writing of Annie Dillard, Fredrick Chr. Brøgger is of the opinion that the writings of Emerson and other American transcendentalists have "strongly anthropomorphic and ultimately anthropocentric connotations" (Brøgger 2007). However, anthropomorphism can also enable humans to relate to nonhuman nature, and therefore, cannot be equated with anthropocentrism (Goralnik and Nelson 2012). Brøgger (2007) also quoted some statements of Emerson from his book *Nature* to prove his point, though he did not mention that Emerson had also said "the greatest delight which the fields and woods minster, is the suggestion of an occult relation between man and the vegetable". Both humans and nature recognize each other's existence and gesture to each other to produce an effect which is similar to "that of a higher thought or a better emotion". However, after pondering over the question of who has "the power to produce this delight", Emerson said that it "does not reside in nature, but in man, or in a harmony of both" (Emerson 1849, 4). It is quite common for many thinkers to debate within their own minds about the correct interpretation of a certain observation or concept, to finally arrive at what they believe to be the truth. Though Emerson initially thought that humans have the power to produce the "delight" of the spiritual relation between humans and nature, he later realized that only a harmonious co-existence and co-evolution of humans and nature could produce this effect. Here, even if Emerson initially thought about human control and mastery over human-nature relation, he finally moved away from this position to that of a partnership between humans and nature. In this way, Emerson also passed from the material to the transcendental and from exploitation to participation.

Because Emerson had expressed his belief that the true image of nature is essentially poetic, we take a look into some of his poems for their ecocentric images and recognition of intrinsic value in nature. Emerson's poetry could easily identify with nature, visualize its soul, and recognize the bonds of kinship that bind the poet to nature. In the poem *River* – written at a young age – he gazes at the river that he had admired as an infant, and says

> Oh, call not Nature dumb; / These trees and stones are audible to me, / These idle flowers, that tremble in the wind, / I understand their faery syllables, / And all their sad significance. The wind / That rustles down the well-known forest road - / It hath a sound more eloquent than speech.

Emerson recognized the voice of nature all around him, when he wrote "The stream, the trees, the grass, the sighing wind, / All of them utter sounds of 'monishment / And grave parental love". A young Emerson looked at the various natural entities around him as his kins and guardians, and this is explicitly expressed as

> I feel as I were welcome to these trees / After long months of weary wandering, / Acknowledged by their hospitable boughs; / They know me as their son, for side by side, / They were coeval with my ancestors, / Adorned with them my country's primitive times, / And soon may give my dust their funeral shade.
>
> *(Emerson 1867–1911, 422–423)*

These lines reflect the realization of Emerson about the seemingly eternal presence of nature, which existed along with his ancestors, and even before that in ancient times. He could also visualize their presence beyond his times, when they would cast their shade over his grave. Emerson also did not hesitate to acknowledge the wondrous abilities of nature, when he followed the many feats of the bee (*The Humble-Bee*), who he said is "wiser far than human seer", and called it a "Yellow-breached philosopher!" It has the wisdom to take all that is "sweet" and substantial and knows how to remain unscathed by the vagaries of nature such as the freezing cold of winter, when humans are tortured by "want and woe" (Emerson 1867–1911, 69–70). The poet had the "inner eye" to comprehend the language of nature "When the pine tosses its cones / To the song of its waterfall tones", and could talk to "birds and trees" (*Woodnotes I*: Emerson 1867–1911, 74). In *Woodnotes II*, nature in the form of a pine tree spoke to the poet (Ziltener 2007). Here, Emerson bowed to the superiority of nature over humans, when the pine tree reminded him that while the poet's creations could only be understood by the Americans, the verses manifested in the tree were universal. Its branches could speak in all languages, and its song could bring together "the world in music strong". Nature also offered to "heal the hurts" and be the companion of the poet, who could mingle with "…. the primal mind / That flows in streams, that breathes in wind" in order to get a transcendental experience of life (Emerson 1867–1911, 78–83). In the later sections of this chapter, we will find similar thoughts finding their place in the poems of Rabindranath Tagore from India, who wrote in Bengali.

Humans place intrinsic value in all human beings, and more so in their kins and near and dear ones. Emerson the poet also found intrinsic value in the "pale violet" which continued to bloom in the autumn cold, when all the other plants withered, and the tall trees sheltering it shed their leaves (*The Violet*). The violets brought the sad memories of his late sister, when he wistfully wrote "I had a sister once who seemed just like a violet"; but "When the violets were in their shrouds, and Summer in its pride, / She laid her hopes at rest, and in the year's rich beauty died" (Emerson 1867–1911, 121). In *Berrying*, the blackberry vines asked him whether he had received any wisdom from their berries (Emerson 1867–1911, 71). In The *Titmouse*, the little titmouse (or the chickadee as some critics have pointed out) gave him courage when he was chilled by the freezing winter miles away from his home and felt helpless and lost. Then with the "saucy note" of a "tiny voice" – "chic-chic-a-dee-de", the little bird appeared to him like "a feathered lord of land" in his "sylvan fort", who "Flew near, with soft wing grazed my hand, / Hopped on the bough, then, darting low, / Prints his small impress on the snow". The tiny bird also taught Emerson that nature's magnificence did not depend on size, and the poet pledged to wear henceforth no stripe other than the ash and black of the "dare-devil" bird. He said, "I think no virtue goes with size; / The reason of all cowardice / Is, that men are overgrown, / And, to be valiant, must come down / To the titmouse dimension" (Emerson 1867–1911, 243). Emerson found a divine presence in nature when in his poem *Good-bye*, he bade good bye to the "proud world" and went back to his "own hearth-stone / Bosomed in yon green hills alone", where green arches echo the blackbird's refrain in "A spot that is sacred to thought and God". When safely ensconced in his "sylvan home", he felt that he can ignore the ancient glory, "the pride of man", and the knowledge of the "learned clans", "For what are they all, in their high conceit, / When man in the bush with God may meet?" (Emerson 1867–1911, 33). Perhaps infused with this spirit of being a part of nature, Emerson asked the botanist to go to his "learned task", while he remained "with the flowers of spring" (*Botanist*: Emerson 1867–1911, 309).

Finally, we come to his poem of over 300 lines titled "*The Adirondacs: A Journal* dedicated to my fellow travellers in August, 1858". This refers to his journey to the Adirondack mountains in New York State, USA, where he was greatly impressed by the tall white pines, which to him had an oracular capacity. Growing on a slope and bent by the breeze to look like a file of climbers, they inspired him to write: "... in the evening twilight's latest red, / Beholding the procession of the pines" (Emerson 1867–1911, 197; Jamieson 1958). The experience of nature at Maple Camp induced Emerson to move more often from reductionism to holism, and from the mundane to the transcendental. For instance, he wrote: "Two doctors in the camp / Dissected the slain deer, weighed the trout's brain, / .... Not less the ambitious botanist sought plants, / Orchis and gentian, fern and long whip-scirpus,...". Then he noticed that

> Above, the eagle flew, the osprey screamed, / The raven croaked, owls hooted, the woodpecker / Loud hammered, and the heron rose in the swamp. / As water poured through hollows of the hills / To feed this wealth of lakes and rivulets, / So nature shed all beauty lavishly / From her redundant horn.
>
> *(Emerson 1867–1911, 197–198)*

The artist William J. Stillman, a leading member of the Boston Club, wrote in his autobiography that the image of Emerson "groping about in the darkness of the primeval forest" – perhaps trying to understand the divine and the sublime in nature – was etched vivid in his memory.

An enamored Emerson wrote in *In Adirondack Lakes*

> Judge with what sweet surprises Nature spoke / To each apart, lifting her lovely shows / To spiritual lessons pointed home, / .... Hark to that petulant chirp! what ails the warbler? / Mark his capricious ways to draw the eye. / Now soar again. What wilt thou, restless bird, / Seeking in that chaste blue a bluer light, / Thirsting in that pure for a purer sky?
>
> *(Emerson 1867–1911, 199)*

Here, Emerson finds a "mystic hint" in the "form and ways" of all creatures, which is "not clearly voiced", but conveys a new realization, and this new knowledge makes him ask the restless warbler what it seeks to find in the blue sky.

However, despite his new-found realizations about nature in the wilderness of the Adirondacks, Emerson finally had to return to the hard realities of the Anthropocene, when he wrote toward the end of his long poem:

> We flee away from cities, but we bring / The best of cities with us, we praise the forest life: / But will we sacrifice our dear-bought lore / Of books and arts and trained experiment, / O no, not we!
>
> *(Emerson 1867–1911, 201)*

Brøgger (2007) commented on the contradiction in Emerson and other transcendentalist writers that on the one hand they pointed out the separateness of humans from nature, while on the other, they expressed "pantheistic sentiments". This gave rise to a paradox. Alan Hodder (Hodder 1989, cited in Wittenburg 2012) was also severe on Emerson and commented sarcastically that Emerson's popular portrayal as a devoted nature worshiper was a caricature. He also dubbed him as an armchair nature lover whose commitment to nature could not be trusted. It is true that on several occasions, Emerson professed the excellence of humans, and yet perhaps as many times or more, he also exalted a transcendental experience of nature, recognizing its equality and even superiority over humans. In several poems, he tried to understand the language of nature in the birds or in the majestic pines of the forest, thereby attributing intrinsic value to nature. The image of Emerson "groping" in the twilight of a dense forest to grasp the underlying meaning of nature is perhaps an appropriate way of describing his perceptions of nature. His thoughts might not have been totally free from their anthropocentric roots, but it cannot be denied that he made frequent forays into the ecocentric domain. The rather scathing criticism of Emerson by Hodder is, therefore, unwarranted and unfair to the transcendentalist thinker. Writing in the post-Aldo Leopold and post Silent Spring era, it is easy to criticize the thinkers who pioneered the concept of reverence for nature and were engaged in the quest of finding intrinsic value in nature. One must remember that they lived in a period in which Cartesian concepts were still overbearing, when new frontiers of seemingly unlimited resources of the Earth were being opened every now and then, and when science and technology were making rapid strides and held the promise of resolving almost every conceivable problem. If not anything else, the poems of Emerson quoted in this chapter establish his love and commitment – even if transient – to the magnificent white pine and the tender violet, the tiny and capricious warbler, the brave titmouse, the blackbird, the "flowers of spring" and many other elements of nature, and most importantly, nature in its entirety as a living and conscious entity.

7.2.2 Henry David Thoreau

Henry David Thoreau (1817–1862) was an American philosopher, poet, and environmental scientist. *Walden* is the major work written by him, which comprises his experiences with nature when he lived in a log cabin on the shore of the Walden Pond in Concord, Massachusetts, USA, from July 1845 to September 1847 (Furtak 2019). Thoreau has been lauded by Ralph Waldo Emerson "as the attorney of the indigenous plants". It has also been said that his writings reflected "perceptions that were pantheistic and animistic", and he had a sympathetic attitude toward paganism. He was deeply interested in Native American worldview, "spirituality and life practices". However, his understanding was ecological in nature, and he recognized the interdependence of all forms of life (Taylor 2012).

In her analysis of two major works of Thoreau, namely, *Walden* and *The Maine Woods*, Nicole Elaine Wittenburg found Thoreau's views as ecocentric, because he accepted nature in its own right, and not because of its usefulness to humans (Wittenburg 2012). While reviewing Lawrence Buell's analysis of Walden and other works (Buell 1995, cited in Wittenburg 2012), she noted that Buell found Thoreau placing nature's interests ahead of human interests, and this viewpoint expressed in Walden had developed over a lifetime of varied experiences with nature, which were marked by a process of change and adoption of newer viewpoints. Thoreau's philosophy was also influenced by precise and scientific observations of nature rather than only abstract transcendentalism (Walls 1995, cited in Wittenburg 2012). Some passages in *Walden* can acquaint the reader with the ecocentric thoughts of Thoreau, which acknowledged nature in its own songs and sounds, and in its own rights of existence. For example, when he described the daily singing bouts of the whippoorwills in the evening, and the screech owls which followed them, their "wailing, their doleful responses" seemed to him to represent the "dark and tearful side of music", and appear like "spirits … of fallen souls that once in human shape night-walked the earth". He thought that "the idiotic and maniacal hooting" of the owl "is a sound admirably suited to swamps and twilight woods … suggesting a vast and undeveloped nature which men have not recognized". He had a similar feeling while listening to the chorus of the bullfrogs in the pond, relaying the sound from one to the other, as if they were seating at a festive table and passing the wine which had distended their flabby bellies (Thoreau 1995, 70–72). It is noteworthy that Thoreau did not attempt to impart any anthropomorphic qualities to these denizens of the forest, and yet made them come alive in their own right, infused as they were with intrinsic value in their existence in nature. In his encounter with nature, Thoreau said that "I go and come with a strange liberty in Nature, a part of herself". He also said: "Sympathy with the fluttering alder and poplar leaves almost takes away my breath". Thoreau was truly ecocentric because his identification with nature extended from the birds to animals like the fox, skunk, and rabbit that came alive at night, to the fallen leaves of the trees, and to the lake, whose "serenity is rippled but not ruffled" (Thoreau 1995, 74). He enjoyed "the friendship of the seasons" and felt "that the most sweet and tender, the most innocent and encouraging society may be found in any natural object, even for the poor misanthrope and most melancholy man". Thoreau did not blame the rain for preventing him from hoeing his bean-field and spoiling his seeds and the potatoes in low-lying areas. Instead, he realized that "it would still be good for the grass on the upland, and, being good for the grass, it would be good for me" (Thoreau 1995, 72). These statements show that Thoreau was not simply an abstract thinker, he had precise ecological knowledge, and his thinking extended to the functioning of the entire forest ecosystem. When asked about his loneliness in his solitary living in the Walden cabin, Thoreau responded that he was no more lonely than the pond, the brook, the sun, the north star, the wind, thaw, or the loon with its raucous laughter, the sorrel, horsefly, bee, "the first spider in a new house", a solitary mullein or dandelion (plants), or a bean leaf. Thus, his identification with nature was all-pervasive, spreading to all its living and non-living elements. He asked: "Shall I not have intelligence with the earth? Am I not partly leaves and vegetable mould myself?" (Thoreau 1995, 75).

Thoreau was also concerned about human exploitation and decimation of the forest around the Walden Pond, and recalled that when he first visited the pond, it was surrounded by a dense growth of tall pines and oaks, with grape vines forming bowers in some of the coves of the pond. However, when he visited the place again, some years after he had departed from the place, the woods were very much depleted, and the villagers planned to develop a piped supply to provide water to the households. Nevertheless, despite being robbed of its woody cover and being fringed by a railroad, the Walden pond struggled to remain unchanged in the purity of its water (Thoreau 1995).

The poems of Thoreau also acquaint the reader with the ecocentric images that he built in his poetry. In *Nature*, he declared that his aim was not to become a meteor or comet high up and dazzling in the sky, but to be a gentle breeze among the reeds by the river, where he wanted to blow gently through a reed growing in a lonely recess, or he could "whisper" through the leaves in the forest. His only desire was to stay close to nature and be her "child / And pupil, in the forest wild, / Than be the king of man elsewhere" (Thoreau 1895, 7). In *To My Brother*, which seems to be dedicated to his brother who died of tetanus, he wondered whether he will find his brother "Along the neighboring brook" or on "…. the brink / Of yonder river's tide?" He also wondered which bird would bring him the word of his brother, and he found "A sadder strain mixed with their song". The finch and the thrush which used to sing for him did not return anymore, because "They have remained to mourn, / Or else forgot" (Thoreau 1895, 23). In *Nature's Child*, he spoke on behalf of the "autumnal sun", which ran a race "With autumn gales", and under whose glow the hazel flowers bloomed and grapes ripened in the "bowers". Yet the latent winter made it sad as it heard "the rustling of the withered leaf" (Thoreau 1895, 30). In this way, nature came alive in many of his poems. In *Stanzas Written at Walden*, Thoreau experienced a feeling of oneness with the joy of the mouse nibbling the hay, the Chicadee's subtle musical notes, the "Fair blossoms" adorning "the cheerful trees", the sound of cricket, "fagot on the earth", or the cracking ice on the Walden Pond (Thoreau 1895, 34). The *Winter Scene* did not portray human activities, but that of almost all possible inhabitants of the woods and riverbank: the rabbit, mouse, ferret, marmot, owlet, raven, squirrel, partridge, and others, and plants such as the willow and the alder with their catkins drooping (Thoreau 1895, 35). He paid his tributes to the crow – the "dusky spirit of the wood", and a "Bird of an ancient brood" – which moved restlessly "From wood to wood, from hill to hill, / Low over forest, field, and rill (Thoreau 1895, 36). He also addressed "... a Stray Fowl" and the "mountains", covering a wide canvas of nature in his poems. An important aspect of winter for Thoreau was that it was the time when "Under the hedge, where drift banks are their screen, / The titmice now pursue their downy dreams" [https://mypoeticside .com/show-classic-poem-31153]. Nature also came to him – not in the role of a father – but as a mother, and he wrote:

> Sometimes a mortal feels in himself Nature / – not his Father but his Mother stirs / within him, and he becomes immortal with her / immortality. From time to time she claims / kindredship with us, and some globule / from her veins steals up into our own. (*I am the Autumnal Sun*) [https:// allpoetry.com/I-Am-The-Autumnal-Sun]

And this kinship with nature may be the most important message from Thoreau to his readers.

Thoreau made an excursion to Mount Ktaadn, the second highest mountain in New England, in 1846. The forests at the foot of the mountain were subject to intensive lumbering operations, and Thoreau observed that the pine tree that stood majestically on the lake shore, its branches swaying in the "four winds", "and every individual needle trembling in the sunlight", now was prostrate and "sold ... to be slit and slit again...". Here, Thoreau spontaneously recognized the existence of the tree in its own right as an individual organism, and not as a piece of lumber meant for human use. He compared the role of humans to "so many busy demons" whose objective was "to drive the forest all out of the country, from every solitary beaver-swamp and mountain-side" (Thoreau 1906, 9). However, he did not totally ignore the use value of trees, when he observed that apple trees in this area could benefit from some grafting exercise. This is a point worth-making, as it shows that to recognize extrinsic value in another entity does not necessarily reflect an anthropocentric bent of mind, but denying intrinsic value to all nonhuman entities does. Thoreau's ecocentric views were expressed time and again, when he described his impressions of the "Maine wilderness", which "is a country full of evergreen trees, of mossy silver birches and watery maples, the ground dotted with insipid small, red berries, and strewn with damp and moss-grown rocks". He had concern for the whole ecosystem, not only its more spectacular inhabitants, and described it as the home of trout, salmon, shad, pickerel, "the chickadee, the blue jay, and the woodpecker", the screaming fish hawk and the eagle, the hooting owl, and howling wolves. It was home to "the moose, the bear, the caribou, the wolf, the beaver, and the Indian". So, humans could also share this life, living in nature, and experience "the inexpressible tenderness and immortal life of the grim forest". Here, nature was forever young, "like a serene infant" (Thoreau 1906, 47). However, Wittenburg (2012) also pointed out the contradictions of Thoreau regarding human treatment of nature, because on the one hand he was in favor of vegetarianism, but on the other, expressed doubts about the ethics of agriculture. At the same time, he wrote in *Walden* that he wanted the sons of his friends to go hunting – "though sportsmen only at first" – as he found that to be "one of the best parts of my education" (Thoreau 1995, 120). However, he was aghast at the cruelty of hunting after watching the hunting of moose in the Maine forests, and wrote about his non-violent, biocentric realization in *The Maine Woods*: "Every creature is better alive than dead, men and moose and pine trees, and he who understands it aright will rather preserve its life than destroy it" (Thoreau 1906, 68).

7.2.3 Walt Whitman

Nature had also made its presence very prominently felt in the poetry of Walt Whitman (1819–1892), who is regarded by many as a humanist forming a bridge between realism and transcendentalism. However, Whatman's humanism is distinct from that of many other humanists who only kept the humans within its ambit. Whitman's humanism was an extended one, in which its recipients included the nonhuman world. When Whitman said that humans return to nature to become nature, to become plants, fishes, and hawks, when he wrote that the disconnect between humans and nature would disappear to allow the two entities to join together, or when he likened the human child to grass – the "babe of the vegetation" – then we experience an expanded "other" in the poetic world of Whitman. George Santayana, the American ecocentric philosopher, thought that among the American writers, Walt Whitman stood out in his rejection of anthropocentrism. Whitman overcame the American "genteel tradition", and meted out a democratic treatment "to the animals, to inanimate nature, to the cosmos as a whole". In this sense, Whitman was a pantheist in his own way (Sessions 1995, 167). Whitman's unique form of pantheism is evident in many of his poems, parts of a few of which are shared and discussed in this section.

In the *Song of Myself*, Whitman pondered over the question posed to him by a child, who asked,

> what is the grass? …. / How could I answer the child? I do not know what it is any more than he. / I guess it must be the flag of my disposition, out of hopeful green stuff woven. / Or I guess it must be the handkerchief of the Lord, / A scented gift, … / Or I guess the grass is itself a child, the produced babe of the vegetation.
>
> *(Whitman 1855–1892, 20)*

In these lines, the profound wonder of Whitman, his quest for the essence of nature, blended with his spiritual realizations, and imparted a distinct intrinsic value to the grass, which appeared to him to be a child, a babe of the vegetation. The grass-child and the human child mingled and blended into each other. The grass imparted in him an eternal presence, and he said, "I bequeath myself to the dirt to grow from the grass I love, / If you want me again look for me under your boot-soles" (Whitman 1855–1892, 53). His identification with nature, and the unity in nature are reflected in the stanza

> The wild gander leads his flock through the cool night, / Ya-honk he says, and sounds it down to me like an invitation, / The pert may suppose it meaningless, but I listening close, / Find its purpose and place up there toward the wintry sky. / The sharp-hoof'd moose of the north, the cat on the house-sill, the chickadee, the prairie-dog, / The litter of the grunting sow as they tug at her tits, / The brood of the turkey-hen and she with her half-spread wings, / I see in them and myself the same old law.

Later, in the same poem he again said: "I believe a leaf of grass is no less than the journey work of the stars" (Whitman 1855–1892, 24, 35). Whitman saw himself and the male and the female, the entire humanity, blend with nature, when he wrote in *I Sing the Body Electric*: "As I see my soul reflected in Nature, / As I see through a mist, one with inexpressible completeness, sanity, beauty, / See the bent head and arms folded over the breast, the Female I see" (Whitman 1855–1892, 57).

Whitman also searched for meaning in the language of nature, when he observed "Oxen that rattle the yoke and chain or halt in the leafy shade, what is that you express in your eyes? / It seems to me more than all the print I have read in my life". He did not want to change the way they are, and the way they communicate, and create music, because he asked

> And do not call the tortoise unworthy because she is not something else, / And the jay in the woods never studied the gamut, [a complete scale of musical notes, or the range of a voice or instrument] yet trills pretty well to me, / And the look of the bay mare shames stillness out of me.
>
> *(Whitman 1855–1892, 57; words within box brackets are author's words)*

Thus, there was profound respect for nonhuman animals in Whitman's perception, and he never tried to interfere with or alter their uniqueness.

In *There Was a Child Went Forth*, Whitman found how the elements of nature that a child saw became parts of him:

> The early lilacs became part of this child, / And grass and white and red morning glories, and white and red clover, and the song of the

phoebe-bird, / And the Third-month lambs and the sow's pink-faint litter, and the mare's foal and the cow's calf, / And the noisy brood of the
barnyard or by the mire of the pond-side, / And the fish suspending themselves so curiously below there, and the beautiful curious liquid, / And the
water-plants with their graceful flat heads, all became part of him.

Here, Whitman saw the child imbibing an assortment of nature without
any hierarchy: beautiful flowers and the discarded weeds, the wild and
the domestic animals, and the fish and aquatic plants and the water itself,
all coming together. He went on to include the crops, the fruit plants, the
weeds, men and women and boys and girls, and the other elements of the
environment and the man-made city as parts of the child:

The field sprouts, ... / ... the esculent roots of the garden, / And the apple-
trees cover'd with blossoms .... and the commonest weeds by the road, /
.... And all the changes of city and country wherever he went. / His own
parents.

Almost everything around the child,

the language, the company. .... / Men and women crowding fast in the
streets, ... / The streets themselves and the façades of houses, ... / Vehicles,
teams, ... / The village on the highland, ... / Shadows, aureola and mist,
.... / The hurrying tumbling waves, .../ The horizon's edge, ... / These
became part of the child who went forth every day, and who now goes,
and will always go forth every day.

All these became integrated into the child (Whitman 1855–1892, 207–208).
Through these all-encompassing lines, Whitman integrated all the elements
of nature – human and nonhuman, natural and man-made, tangible and
intangible – into the existence of the child, and this perpetuated in him, and
in all humans. In fact, these and many other similar writings of Whitman
can be said to form a poetic basis for the first law of ecology put forward by
Barry Commoner, which states that "Everything is connected to everything
else" (Commoner 1971, 35). Just as ecology recognizes connections between
all entities in the ecosystems and the entire biosphere, "ecological spirituality" involves the connections of humans with all the elements that surround
them and permeate their selves.

Donald P. St. John interpreted Whitman's spirituality as "ecological spirituality", whose essence is its "identification with the other". There is the
"greatest joy" in this identification, which gives birth to the "most dramatic
poetry" of Whitman. The "cosmic self" is realized through "transcending
ego and becoming the other". The myriad living beings, forms, and objects

that the child "sees" become a part of him, so that his own self unites with that of the others and makes them its own. This "bipolar movement" personalizes "the macrocosm" to enrich "the microcosm". This "expanding and sympathetic self is natural to humans", and it is most vividly present in the child, but is later distorted to become confined to a narrow, individual self (St. John 1992). From this realization, we may surmise that anthropocentrism is the evident outcome of this narrowing of the human perspective, which first gets confined to a species, and then to some specific hierarchical criteria, which could be gender, color or creed, power, wealth, social status, and so on. Nature is mostly placed in the lowest rung of this ladder, and is, therefore, deprived and exploited the most. M. Jimmie Killingsworth quoted the lines in *Song of Myself* "Voices of the interminable generations of prisoners and slaves, / ... Of the deform'd, trivial, flat, foolish, despised, / Fog in the air, beetles rolling balls of dung" (Killingsworth 2010). In the words of Killingsworth, Whitman "flattened [the] hierarchy" of the "great chain of being". His humanism was spread far and wide to make him an ecocentric poet in the real sense, and his *Song of Myself* transcended into a "Song of Everybody". We also find that Whitman was not an advocate of renunciation, because he liked to enjoy the gifts of the Earth, but only to the extent to leave enough for all the others, and to see all the others – big and small – in their cherished places. One is reminded of one of the best-known quotes of Gandhi: "The world has enough for everyone's needs, but not everyone's greed". This is also reflected in the following lines of Whitman:

I resist any thing better than my own diversity, / Breathe the air but leave plenty after me, / And am not stuck up, and am in my place. / (The moth and the fish-eggs are in their place, / The bright suns I see and the dark suns I cannot see are in their place, / The palpable is in its place and the impalpable is in its place).

(Whitman 1855–1892, 207–208)

St. John also contended that a poet "must be able to identify with both sexes", and in the poem "We Two, How Long We Were Fool'd" (Whitman 1855–1892, 63–64), Whitman portrayed the perception of nature "by the new Adam and New Eve", where "conventional sexual stereotypes are transcended by identification of the lovers with the larger context and energies of the earth" (St. John 1992). In this poem, Whitman wrote:

We two, how long we were fool'd, / Now transmuted, we swiftly escape as Nature escapes, / We are Nature, long have we been absent, but now we return, / We become plants, trunks, foliage, roots, bark, / We are bedded in the ground, we are rocks, / We are oaks, we grow in the openings side by side, / We browse, we are two among the wild herds spontaneous as any, / We are two fishes swimming in the sea together, / We are what locust blossoms are, we drop scent around lanes mornings and evenings, / We are also the coarse smut of beasts, vegetables, minerals, / We are two predatory hawks, we soar above and look down, / ... We are snow, rain, cold, darkness, we are each product and influence of the globe, / We have circled and circled till we have arrived home again, we two, / We have voided all but freedom and all but our own joy.

It can also be said that though the poem depicts male-female or even male-male interactions, using the various images of nature as metaphors, it also gives a "clear ecocentric message", because the "two" transmute into nature in the form of plants, animals, entire ecosystems, as well as the abiotic components of nature (Gupta 2022). St. John interpreted this poem as the transcendence of the lovers beyond "conventional sexual stereotypes" to identify with a "larger context and energies of the earth". In *Passage to India*, Whitman sang to the "great achievements of the present", the "works of engineers, / Our modern wonders,..." the feats of technology at the Suez Canal and the railroads, but also "the infinite greatness of the past!" However, he also said "Nature and man shall be disjoin'd and diffused no more, / The true son of God shall absolutely fuse them" (Whitman 1855–1892, 232–235). Did Whitman mean that the poet is "The true son of God", who will unite Europe and Asia, the east and the west? (Gupta 2022). After all, Whitman saw the poet as "the caresser of life", whose poetry springs from the Earth's poetry. It is the Earth which generates the primary words, and sometimes "speaks" through its own language. It is for the poet to comprehend their meaning (St. John 1992). To Whitman, men, women, their bodies, their love, all were fused with nature, when he said in *Spontaneous Me*:

Spontaneous me, Nature, / The loving day, the mounting sun, the friend I am happy with / The arm of my friend hanging idly over my shoulder, / The hillside whiten'd with blossoms of the mountain ash, / The same late in autumn, the hues of red, yellow, drab, purple, and light and dark green, / The rich coverlet of the grass, animals and birds, the private untrimm'd bank, the primitive apples, the pebble-stones, / .... The real poems, (what we call poems being merely pictures,) / .... The body of my love, the body of the woman I love, the body of the man, the body of the earth, / ... The hairy wild-bee that murmurs and hankers up and down, that gripes the / full-grown lady flower ....

In this way he continued to describe the love of humans, the love of the Earth, and their bodies, their sexuality, their procreation, and "The merriment of the twin babes that crawl over the grass in the sun .../ The continence of vegetables, birds animals", which came spontaneously together without any wall separating them (Whitman 1855–1892, 61). Many meanings may be made of these words, but we can always find the presence of nature in all its diversity mingling with its procreations, including humans.

Besides his poetry, nature also deeply permeated his prose writings. Whitman dedicated

> the last half of these Specimen Days to the: bees, black-birds, dragon-flies, pond-turtles, mulleins, tansy, peppermint, moths, (great and little, some splendid fellows,) mosquitoes, butterflies, wasps and hornets, cat-birds, (and all other birds,) glow-worms, (swarming millions of them indescribably strange and beautiful at night over the pond and creek,) water-snakes, crows, millers, cedars, tulip-trees, (and all other trees,) and to the spots and memories of those days, and the creek.
>
> *(Whitman 2005, 86)*

By the "creek", Whitman meant the Timber Creek in New Jersey, where he was recuperating from his illness (St. John 1992).

In "New themes entered upon", Whitman said that after one had exhausted everything in business, politics, love and friendship, nature remained and it brought out the bond of humans "with the open air, the trees, fields, the changes of seasons ...". Nature could bring out "from their torpid recesses, the affinities of a man or woman with the open air, the trees, fields, ...". After the "long strains of war, and its wound and death" (American Civil War: 1861–1865), his strolls took him to a ravine filled with "bushes, trees, grass, a group of willows, ... a spring of delicious water ..." and he wrote, "Never before did I get so close to nature, never before did she come so close to me".

Whitman made elaborate notes on the songs, activities and "evening frolic" of cat-birds, thrushes, kingfishers, quails, and other birds, the majesty of "the old, warty, venerable oak", and all the assorted scents and sights of nature around him. In the *Lesson of a Tree*, he gazed with wonder at "a fine, yellow poplar, ... strong, vital, enduring! How dumbly eloquent!" It was a symbol of "importability and being... Then the qualities, almost emotional, palpably artistic, heroic, of a tree; so innocent and harmless, yet so savage" (Whitman 2005, 58, 62–63, 73).

Both St. John and Emily Gerhardt acknowledged Whitman's attention to nature's "minutiae", and his remarkable "humility" to all the creations of nature (St. John 1992; Gerhardt 2014, 73). Whitman's humility was extended to the smallest creature. This perception is reflected in the following stanza of his poem *Song of Myself*:

> I believe a leaf of grass is no less than the journey work of the stars, /
> And the pismire is equally perfect, and a grain of sand, and the egg of

the wren, / And the tree-toad is a chef-d'oeuvre for the highest, / And the running blackberry would adorn the parlors of heaven.

(Whitman 1855–1892, 35)

In this humility, the equal regard to all the creations of nature, and his feeling of oneness with them, lie the essence of the ecocentric spirit of Walt Whitman.

7.2.4 Emily Dickinson

In contrast to Emerson, who made occasional journeys into wild nature and searched for its higher meaning, and Thoreau, who lived for some time amidst nature in his *Walden*, the encounters of Emily Dickinson (1830–1886) with nature (at least as reflected in her poetry) was almost entirely confined to her garden, its flowers and trees, insects, and the winged visitors who happened to come there. In the preface to the first series of the collection of Emily Dickinson's poems (Dickinson 2004, 2), Thomas Wentworth Higginson expressed his awe at the quality of her poems, which reminded one of the poems of William Blake, with "flashes of wholly original and profound insight into nature and life". However, since the selection of poems was published after her death, the titles of most of the poems were assigned by the compilers or editors, and "in many cases these verses will seem to the reader like poetry torn up by the roots, with rain and dew and earth still clinging to them, giving a freshness and a fragrance not otherwise to be conveyed". Emily Dickinson lived a life in which her conversations with "nature" did not require her to visit the wilderness of Emerson and Thoreau. She met nature in her own garden in the flowers and the bees and butterflies that frequented them. The common robin was the bird that found its place in her soul. In the preface to the second series of her poems, Mabel Loomis Todd – who was a friend – wrote in 1891:

> Storm, wind, the wild March sky, sunsets and dawns; the birds and bees, butterflies and flowers of her garden, with a few trusted human friends, were sufficient companionship. The coming of the first robin was a jubilee beyond crowning of monarch or birthday of pope; the first red leaf hurrying through *"the altered air"*, an epoch.
>
> *(Dickinson 2004, 39)*

Christine Gerhardt, in her study on Emily Dickinson's and Walt Whitman's poetry, observed that these two poets had never undermined or ignored the value of nature toward serving as a habitat or home for harboring and sustaining life. About four hundred poems of Dickinson mention flowers or parts of flowers, along with small birds like robin, oriole, blue jay, bees, insects, and even snakes finding their place in many poems. These small creatures hardly found their place in the "genteel imagination" of those times, and their frequent appearance in Dickinson's poems marked a "passionate gesture" toward small nature. Gerhardt also drew our attention to the considerable botanical/ecological knowledge that Dickinson had, which was a characteristic of her times. It was thought necessary to know about the small and apparently insignificant life forms in order to better understand and appreciate the "complex web of life", though Dickinson's interest in "small nature" could partly have stemmed from the "Romantic interest in the ugly or abject" (Gerhardt 2014, 36).

Dickinson (2004, 4) felt that if she could save even one life, her own life would not be in vain, and she wrote: "Or help one fainting robin / Unto his nest again. / I shall not live in vain". We can imagine here that the robin – a small bird – may represent exploited nature, or a woman oppressed by patriarchy, or simply "a lost human being" (Dawoud 2017). Whatever it might be, whether the robin actually represented a hurt small bird, or nature as a whole, or oppressed humanity, or a woman, one might say that they blended and mingled into each other, losing their distinction and the notion of superiority or inferiority, and the unity of all life as well as inanimate nature emerged through the image of a small wounded bird.

Another important feature in Dickinson's nature poems was her attention to "physiological detail" of "flowers, birds, and similar phenomena". She also made keen observations on the structure and behavior of the birds, bees, other insects, and even a snake that visited her garden. These examples point out toward her "proto-ecological literacy" where nature was also valued for its actual "physical existence", and "not just as a route to transcendence" (Gerhardt 2014, 37).

Dickinson's poems are replete with personification of small nature. However, she never used personification with an anthropocentric attitude of control over nature. Instead, her personifications depicted "human-nature relationships that include ethical considerations" (Gerhardt 2014, 39). In many of her poems, we come across an ecocentric recognition of small nature, where they emerge as beings that exist in their own right as conscious entities. A few examples may help illustrate this further.

"May-flower" – also titled as "Arbutus" – is one of the well-known nature poems of Emily Dickinson. Gerhardt (2014, 39) characterized this poem as marked by "intense anthropomorphism". The poem describes the may-flower as "pink, small, and punctual / Aromatic, low", which is "Dear to the moss" and is "Next to the robin / in every human soul". In another poem, Dickinson said "The bee is not afraid of me / I know the butterfly". She also found that the brooks and the breeze were overjoyed in her presence, showing her mental kinship with nature. We can see here that the bee, butterfly, brook or breeze do not become humans, but the poet is proud to know them. Thus, there is no anthropocentric attitude of having superior knowledge or position as a human in the anthropomorphism that we come across in Dickinson's poems. In fact, anthropomorphizing nature – which represents attributing human form or personality to nonhumans – can be seen in many poems of Dickinson. Usually, children do this more often than adults, though many people may carry this trend into their adult-hood. Among the many functions of anthropomorphizing, it could enable humans to connect better with nonhuman nature, and make it easier to connect and feel empathy with nature. Anthropomorphizing may also pro-mote ethical thoughts and make people averse to harming plants and ani-mals, and damaging nature in any form. In one of her many nature poems, Dickinson connected with a bee that was "drunk" with the nectar of the foxglove flower and was "thrown" out just like a drunk tenant was evicted by a landlord. The bee was drunk just like the poet who was inebriated with the balmy summer air and the night dew, whose effects could not be equaled by any wine brewed by humans (Dickinson 2004, 7–8). The con-siderable knowledge that Dickinson possessed on flowers and gardening is also reflected in this poem because the brightly colored foxgloves with their sturdy, broad, and tubular flowers are known to attract bees and bumble-bees in large numbers by virtue of their sweet nectar and fragrant pollen. In "Summer's Armies" (Dickinson 2004, 21), the waking up of the different forms of life such as the "dreamy butterflies", marching "Baronial bees", robins as numerous as the snowflakes of the bygone winter, the orchid that "binds her feather" for her "old lover, Don the sun ..." represents personifi-cation yet no human control. The inanimate nature also came to life, when "Lethargic pools resume the whir / of last year's sundered tune". In her personification of the purple clover, Dickinson described the qualities of this plant that appears early in the season to compete alongside the grass. Its flowers were favored by the bees, butterflies, and hummingbirds alike. It had beauty, because "Her face is rounder than the moon", and was more reddish than the "gown of orchis .... Or rhododendron" (Dickinson 2004, 23). She was brave and persisted against the onslaught of winter. Through this personification, Dickinson showed her great love and appreciation for the tiny flower, which was also "loved" by insects and birds as well, and she made no attempt to create a hierarchy between humans and nonhumans. She saw magnanimity and forgiveness in nature, when she wrote: "Nature, the gentlest mother, / Impatient of no child, / The feeblest or the wayward-est, - / Her admonition mild". The "rampant squirrel" or the "too impetu-ous bird" received her affection, and "Her voice among the isles / Incites the timid prayer / Of the minutest cricket, / The most unworthy flower" (*Mother Nature*: Dickinson 2004, 59).

Gerhardt (2014, 53) contended that when Dickinson identified with small nonhumans, she also "de-centers" her "subject position" and totally relinquished the utilitarian way of viewing nature. This was evident in the poem "The grass", where the grass "which so little has to do – / A sphere of simple green – " had its own identity as a habitat that had butterflies and bees as companions. It swayed in the breeze, and collected the pearly dew at night, having a life of its own.

Dickinson's interactions with nature can be said to create a "dialogue" between her and apparently innocuous plants and animals. She had considerable scientific knowledge of flowers and small animals, though she never flaunted it to show any superiority. We are reminded of the words of Martin Buber – the German philosopher – who gave the example of the various ways that a person could perceive a tree – as a beautiful picture (an object of art), as a distinct biological species, or in terms of its physiology of movement of sap, its respiration, or even strip it of its identity as a being and view it "as an expression of law", or in its "pure numerical relation". In all these perceptions, "the tree remains an object" and is in the domain of "It". On the other hand, one may also perceive it in such a way that makes one "bound up in relation to it". When this happens, the tree no longer remains as "It", but transcends to enter the realm of "Thou" (Buber 1923, 14). In other words, we might say that this transformation represented an ecocentric perception of nature, where the human and nonhuman domains merged into each other.

7.3 Poets and Writers from India

7.3.1 Rabindranath Tagore

Rabindranath Tagore (1861–1941) was an Indian polymath and the first non-European recipient of the Nobel Prize for Literature in 1913. Love and compassion in Tagore's writings not only embraced humans of every caste, creed, and religion, it transcended beyond the human realm to include animate as well as inanimate nature. In *The Religion of Man*, Tagore said that "[Man] finds his own larger and truer self in his wide human relationship. … his multi-personal humanity is immortal". Tagore had integrated into his own life the words of Buddha, who said "Like a mother maintaining her only son with her own life, keep thy immeasurable loving thought for all creatures" (Tagore 1931, 15). These words had guided Tagore in his philosophy of life, as he wrote,

In the world's audience hall, the simple blade of grass sits on the same carpet with the sunbeam and the stars of midnight. / Thus my songs share their seats in the heart of the world with the music of the clouds and forests.

His compassion for "Mother Earth" is expressed in the lines

Infinite wealth is not yours, my patient and dusky mother dust! / .... I will pour my songs into your mute heart, and my love into your love. / .... I have seen your tender face and I love your mournful dust, Mother Earth.

(Tagore 1994, 121)

Kalyan Sen Gupta in his book *The Philosophy of Rabindranath Tagore* (Sen Gupta 2005) observed that Tagore's concerns were not confined to the harmony of human relationships in society. They also strove to achieve people's harmony with nature, which would enable the former to emerge from their self-centered, anthropocentric existence and live a life of spiritual communion with nature. Sukanta Chaudhuri recounted an incident, when after a hectic tour of South East Asia, a somewhat homesick Tagore was revived by the singing of a *koel* in Penang, Malaysia. The frantic notes of the *koel* transported him to his childhood memory of hearing the bird in a country house in India. "In an instant, the foreign setting is absorbed into a familiar landscape of the mind". Chaudhuri (2011) suggested that this lifting of the mind of Tagore partly owed it to his feeling of oneness with nature, and partly to the association of this bird with "Indian love poetry since classical times".

Tagore regarded plants as his "mute friends" who raised their arms toward the sky in their "love of light". He called their language the "primal language of life" that struck the basic strings of our soul. Their messages had no "explicit meaning", yet they reverberated in "melodious notes of eons". In his *Hymn to the Tree (brikshabandana)*, he hailed the trees for creating the "first abode of music" in their branches for the "restless breeze", lending a voice to a voiceless world devoid of the festive notes of the changing seasons. In a poem in which Tagore paid his tribute to his friend Jagadish Chandra (Sir J.C. Bose: Indian scientist), his adulations for Bose were not so much for Bose's innovations in wireless communication and other fields in Physics, but for "bringing to light the first 'silent' voices that arose and spread through grassy meadows to forests", and for hearing the "silent invocations to the sun in the morning breeze". He then wrote,

you could feel the inner agony of the forest; the cry of a mute life surging through countless sprouting seeds, swaying eager branches, fluttering leaves, and meandering roots in their struggle between life and death, that stirred the earth-mother's bosom.

Even when the scientific community was averse to recognize Jagadish Chandra's findings on cell-to-cell communication and neural transmission in plants (though it acknowledged his contributions in wireless communication), Tagore had unflinching faith in the momentous revelations of his friend in unearthing the "language of plants" (Tagore 1982, 847–878) (see

Chapter 3 for the current recognition of Sir J.C. Bose as the "father of plant neurobiology").

Tagore's nature poetry can also be said to reflect his "ecological spirituality", which has its roots in the ancient Indian texts such as the *Vedas* and the *Upanishads*, mingled with his own experiences with the organic life around him in the gardens, forests, wayside fields where nothing but a few wildflowers bloomed now and then, in the dancing ripples on the vast expanse of *Padma*, where he spent his days in a boat, and in countless other places. Sitting in a hotel in Vienna, he dreamed of the trees and creepers in *Santiniketan*, an institution that he had founded in India, and he longed to go back to nature, to return from cacophony to mellifluous music. Nature never ceased to be present in his thoughts and creations. Even the "use value" of trees to Tagore was quite unique: in his stressed moments, he wanted to sit quietly near them, to "bathe his soul every day in the cascade of their 'silent music', which made him feel rejuvenated to enter the world of joy". Nature could come to his life in all situations, anywhere. A *kurchi* tree (*Holarrhena antidysenterica*; common English name: ivory tree, Easter tree, or bitter oleander) that had burst into flowers in all its strength to herald the arrival of spring, despite standing alone against a man-made wall amidst the din and bustle of rushing trains, hordes of bullock-carts, and dusty air of a railway station, could inspire Tagore to pay his tribute to it through a poem. On another occasion, an unknown flower in a country abroad consoled him by urging him to regard the land where it grew his own too ("*Bideshe achena phul* ..."). Many such poems were dedicated to trees, flowers, and nature as a whole, which never ceased to make him wonder at its marvels, and stirred his creative core. Recognition of the intrinsic value of nature came to Tagore spontaneously, upwelling from deep within. He asked the young trees of *Santiniketan*, for whom he had the same love and affection as that for its young residents, to remain as neighbors and friends, to serve as a refuge for the singing birds, to cover the gravelly road with flowers, and to mingle their fragrance with the songs and music of the place. He exhorted them to defy death, to overcome all odds to send their message to the rest of the world (Tagore 1983, 847–878). He was also aware of the ecological cycle of death, decomposition, and emergence of new life. He wrote in his poem *Banaspati* (the tree) about the tree returning its old leaves to the soil, which received it as its earthly treasure to provide nourishment to new saplings (Tagore 1983, 294–295).

Human-nature interactions found their place in Tagore's entire oeuvre: in his songs, short stories, and plays. I cite here just a few of them.

Tagore's vast repertoire of over 2000 songs (*rabindrasangeet* or "Tagore-songs") is replete with ecocentric thoughts and images. In these songs, Tagore sometimes engaged in conversation with nature, or nature opened up its mind to him, or he simply recorded the (silent) conversations between elements of nature. These songs of Tagore can be said to enrich ecocentrism "by adding newer dimensions to its framework and new layers to its core concepts" (Gupta 2016). In one such song, Tagore wrote that when he trained his ears "upon the murmur of the river" with a "heart quivering at the rustling of leaves", he felt that he would be able to "trace the path that goes beyond the evening star" by unraveling "the words of flowers" (Song: *tui phele eshechhish kare*). Here, Tagore takes recourse to nature to guide him along the lost "road that goes beyond the evening stars". In another song, he said that the roadside flowers would send subtle hints through their fragrance to guide him even if his light blew out while traveling along a lonely road (Song: *e pathe ami je gechhi baar baar*). In fact, these songs transport us to Buber's "spheres in which the world of relation arises". Tagore appears to move at ease in this twilight world on the wings of his songs that bridge the gap between silent nature and eloquent humankind. In several other songs, Tagore "rests his faith ... on another form of life or even on the non-living manifestations of nature" (Gupta 2016). For example, he would like to leave the legacy of his songs by splitting them "among the many-hued splendor of flowers", and blend the gliding notes (*meed*) "along the gold-lined edges of clouds". Some would adorn young lovers' garlands, and some "strewn among the *bakul*-bedecked grasses of a forest" (*bakul* or Spanish cherry / Medlar: *Mimusops elengi*), to be picked up by a "carefree wanderer" (Song: *asha jaoar pather dhare*). It is remarkable that Tagore wanted to leave the legacy of his songs to nature, showing his immense faith in and respect for it. He also had the rare sensitivity to interpret the response of the forest to the subtle message of spring, where the flower-buds hear the call of spring in the Cicada-resonant depth of the woods, and ask the poet to be adorned with wreaths and play his flute to be a part of this exchange (Song: *aaj ki tahar barota pelo re kishalay*). Here the poet no longer "remains as an outsider, because he is no longer a master or even a guiding steward possessing superior knowledge... [and] participates in the forest's celebrations of spring as an equal partner" (Gupta 2016). Tagore "records the conversation between a rushing torrent and the stationary *campa* tree", where the rushing movement of the river is contrasted by the silent, unobtrusive journey of the *campa* tree toward light through its "sprouting of young leaves ... [and] cascades of flowers", which is known only to the sky and "the silent stars of night" (Song: *Ogo nodi apon bege pagolpara*). In another song, "the longings of a lover's mind get mingled with the signs and symbols in nature" when the "tremulous *malati*" (jasmine) conveyed the message of the impending visit of the beloved (Song: *aaji godhulilagone ei badologagone*) (Gupta 2016).

In his short story *Subha*, Tagore recounted the story of *Subha* – a speech-impaired girl – who believes that she has a "cursed existence" and engages in silent communication with her two cows and the other animals around her.

She regularly visits the bank of the small river nearby, and tries to find her inner peace by listening to "the sounds of water, the chirping of birds, and all the other voices of nature…". In the stillness of afternoons when the world is devoid of all sound, nature and Subha used to sit face to face, in total silence (Tagore 1950, 204–211). It appears as if nature makes up for her lack of words and comes alive to speak on her behalf, while "the girl is transformed into a part of nature" (Gupta 2023, 89–90). Gupta (2023) cited the philosopher Martin Buber who said that a conversation could be had without a sound or "even a gesture" (Buber 1947, 3), and human relations with nature remained "beneath the level of speech". However, when humans identified themselves with nature, then "words cling to the threshold of speech" (Buber 1923, 13). In another short story *Balai*, a young boy (*Balai*) who had lost his mother in his childhood and was brought up by his uncle and aunt, felt more at ease in the company of plants, whose language he seemed to comprehend, and treated them as his near and dear ones. *Balai* had befriended the sapling of a *shimul* (silk-cotton) tree and entreated his stern uncle not to cut it down. However, when *Balai* left home to pursue his studies, his uncle got this young tree felled, ignoring the protests of his wife, whose filial love for *Balai* was transmitted to the young tree, and who was shocked with grief after its removal. The ecocentric outlook of *Balai* and his aunt was overpowered by the anthropocentrism of the uncle, who considered this tree as a nuisance, because it had no use value to him (Tagore 1990, 405–408).

I cite two plays of Tagore: *Raktakarabi* (The Red Oleander) and *Muktadhara* (The waterfall or "the free stream") to showcase his ecocentric ideals that traveled along different planes. In *Raktakarabi*, *Bishu pagol* (crazy *Bishu*), a tramp, shares his sorrow with the moon to whom he speaks about the "tide of tears reaching the sea of grief" (Song: *o chand, chokher jaler laglo jowar dukher parabare*). Gupta (2016) suggested that when Bishu feels helpless against the king who digs up the bowels of the Earth to mine for gold, coerces the workers, and aims to shackle the Earth itself, he takes recourse to nature (represented by the moon) with whom he feels a deep kinship to confide his sad feelings. The young protagonist *Nandini* is struck by both awe and fear at the immense powers of the king and tries to convince him to free nature from bondage. The red oleander symbolizes her rebel spirit and the yearnings of the Earth to be healed of the wounds inflicted in the maws of the gigantic machines. Her partner, the elusive Ranjan, who remains in the background, is the spirit of struggle personified. The play has been interpreted as the conflict between the agrarian and the industrial civilizations, with the former getting vanquished and subjugated by the latter. However, Tagore left a ray of hope when in the end, the king breaks himself free from the chains of his own ruthless system to turn against the forces governing the machines and joins Nandini in liberating the spirit of the red oleander (Tagore 1985, 191–234).

In *Muktadhara*, Tagore introduces us to *Yantraraj Bibhuti* (Bibhuti the master of machines), who has been able to dam the torrential waters of an unruly stream with his technological prowess. Opposed to him in his worldview is *Dhananjay Bairagi* (Dhananjay the mendicant), who wants to liberate the free flow of the stream, the damming of which represents the tyranny of the king over the people in the downstream areas and human incarceration of nature. In the demonic efforts of *Bibhuti*, we find the use of technology for "enframing", as it challenges nature by regarding it as a "standing reserve", excluding all other qualities that nature may possess. This is contrary to the true, benign, essence of technology, which is basically "revealing" that "brings forth what is unraveled" (Heidegger 1977). *Dhananjay* opposes "such enslaving, exploitative uses of nature", and makes the people of the villages aware of the evils of this despotic attitude toward nature. He is supported by prince Abhijit, who also opposes the abuse of nature by his father, the king, and *Bibhuti*. *Dhananjay* opts for the path of active resistance when he sings "O fire, my brother, I sing victory to you / you are the bright red image of fearful freedom" (song: "*Agun aamar bhai, aami tomari jai gai*" – translated by Tagore himself) as the prison where Abhijit is imprisoned is set on fire. The fire sets Abhijit free, who sacrifices his life while demolishing the dam to restore the free flow of the stream.

These two plays of Tagore would appear like dire prophesies of the present times that confront the threats of climate change, pollution, loss of biodiversity, and dwindling resources, and which remind us of the fourth law of ecology laid down long ago by Barry Commoner (Commoner 1971): "there is no such thing as a free lunch" (Gupta 2016).

7.3.2 Jibanananda Das

Jibanananda Das (1899–1954) is a Bengali poet who is thought to have ushered in modernism in Bengali poetry. Buddhadeb Basu / Buddhadeva Bose (1908–1974) – a noted poet, writer of short stories, novels, plays and essays, and a literary critic – called Jibanananda as "the first real modernist in Bengali poetry" (Basu 1957). Bose characterized Jibanananda's world as "one of tangled shadows and crooked waters, of the mouse, the owl and the bat, of deer playing in moonlit forests, of dawn and darkness, of ice-cold sea-nymphs and the great sweet sea". The poet had close affinity for "all things wanton and non-human"; and "some of his most characteristic poems are on birds and beasts". Bose asserted that "All poets, in a sense, are poets of nature", but Jibanananda's love for nature is very distinct, because he immerses himself in "nature, physical nature". Thus, he worshiped nature, but not like a "Platonist or pantheist". On the contrary, he was like a "pagan" with a "sensuous love" for nature, which neither appeared to him as "tokens or symbols, nor as patterns of perfection, but simply because they are what they are". He was not content with merely seeing nature, but tried to perceive it with his "sense of touch and smell" (Bose 1948, 57–58). In his poem *ghash* (grass), for example, Jibanananda found the deer tearing fragrant green grass the color of tender *batabi* (pomelo: *Citrus maxima*), and he wanted to "drink this fragrance like glass after glass of green wine"; and "cuddle this grass, rub it against my eyes, / feel my feathers in its wings". He would like to be born "as a grass inside a grass ... / descending down the delectable darkness of the body of a dense grass-mother". In *ami jodi hotam* (If I were...), he wondered how he would feel if he became a swan, which lived with its partner in the reed beds of the *Jalshiri* river near a paddy field. On a moonlit night in the month of *falgun* (spring) they would have soared among the "silver harvests" of the sky. Then the harsh reality, the cruelty of man appalled him, when he could hear repeated gunshots that brought them down, silenced them, and at last brought them peace. At this juncture in the poem, Jibanananda merged the life of the swans with that of humans, when he said that such a sudden death might have spared them the pain of "many pieces / fragments of death" that always kept striking them. Nature recurrently came to him in so many of his poems, at special moments of his life. In *kuri bochhor pore* (after 20 years), he imagined that if he met "her" after a gap of 20 years, it would be near sheaves of paddy in the month of *kartik* (autumn), when "evening ravens return to their nest" and the "yellow river mellows with reeds inside the field". Jibanananda could easily empathize with nature, and in *hai chil* (O my eagle), he asked the "golden-winged *chil*" not to fly crying above the *dhansiri* river, because it reminded the poet of the "cane-fruit-like" melancholy eyes of the woman he loved, but who like all beautiful women had gone far away. The persistent cry of the eagle brought back this painful memory. Jibanananda despised the cruelty of human disregard for the sanctity of nonhuman life, when in *shikar* (the hunt) he recounted how the beautiful brown doe, who after a meal of "green pomelo-scented tender grass" waded into the cool bosom of a stream, only to be shot at, and get transformed into warm, red venison on a camp table, accompanied by boastful tales of human prowess (Das 1942). Jibanananda had boundless love for Bengal with its plants, animals, waters, and its entire landscape. In *abar ashibo phire* (I shall come back again), he would like to come back to Bengal "on the bank of *dhansiri*" (probably an imaginary river). He knows he may not return as a human, but in the garb of a "*shankhachil* or *shalik*" (brahminy kite and common mayna, respectively), or "float through the mist" in the form of a "dawn's raven" to rest in the "shade of a jackfruit tree". He might also return as the "pet duck of a young girl with a *ghungur* (small metallic bells strung together) tied around his red ankles", and spend his time floating in the "*kalmi*-redolent" (water spinach plant) waters. But certainly, he would come back "loving the rivers, fields and croplands of Bengal" ... "the green, melancholy land of Bengal moist with the waves of *Jalangi*" (a branch of River Ganga in West Bengal). In *tomra jekhane shadh* (you go wherever you would like to), he wanted to remain in the land of Bengal, where he could

> gaze at the leaves of jackfruit tree shedding in the morning breeze; / the brown wings of *shalik* feel the evening chill / as he hops in the darkness on his yellow legs under white feathers / and then the *hijol* tree calls him close to his heart.

In the same poem, he saw a melancholy woman, who has been born from the *kalmi* creepers on the bank of the pond disappearing in the evening mist, though he knows he would never lose her, because she has remained in the land of Bengal. In *Banglar mukh ami* (I have looked at the face of Bengal), Jibanananda is happy not to wander to find the beauty of the Earth, because he could find it in the face of Bengal. He saw the morning *doel* (oriental magpie-robin) perched under a broad fig leaf, and the piles of fallen leaves from the trees of Bengal – trees standing since eternity (Das 2010). The Bengal of Jibanananda as depicted in these poems is not Bengal as a political unit or units, but a personal haven, where the poet's life is mingled with its everyday nature – its common herbs, trees and bushes, its birds, ponds and rivers, and its melancholy, elusive women – which keep him rooted to the green splendor of this riverine land.

According to Buddhadeb Basu (Buddhadeva Bose), Jibanananda is a perfect example of "a contemplative poet who hears divine music in the *deodar* in the midst of the clamor of consumerism in these frantic, chaotic times" (Basu 1957). Though he did not say it in so many words, these lines of Basu touched upon one of the core philosophies in Jibanananda's poetry, which was a spontaneous recognition of intrinsic value in nature.

7.4 Conclusions

This chapter is meant to show that an ecocentric philosophy is spontaneously adhered to – consciously or subconsciously – by poets and writers belonging to diverse cultures, whose "enchanted" vision of nature could visualize the plants and animals in their joys and sorrows, love and desire just as in humans. However, it is also a fact that while poetic or artistic recognition of intrinsic value in nature is accepted and admired by everybody, adoption of ecocentric principles is still rare in developmental planning such as damming a river to change its flow regime and inundate large areas of forest with stagnant water, excavating a mine, or building a township or industry. The recognition of these principles is even not widespread in conservation, for example, while creating a nature reserve, or restoring the ecosystem services of a given area. This disconnect between the poets, writers, and creative artists, and the policy makers need to be removed from the worldview of humans. The role of Earth jurisprudence in initiating this move has been discussed in Chapter 9. An approach for a more ecocentric future could be to introduce ecocentric creative writing and art in the formal and informal education content meant for children and youth, so that these could shape their *weltanschauung* and make them more sensitive and accommodative to nature's needs.

8

BIOSPHERE, NOÖSPHERE, AND TECHNOSPHERE

8.1 Introduction

The term biosphere, that is, the sphere of life, was first proposed by the Austrian geochemist Eduard Suess (1831–1914) in 1875. He also proposed two other "spheres" – the hydrosphere or the sphere of water, and lithosphere or the sphere of rock, that is, the Earth's crust. Another sphere – the atmosphere or the sphere of air – was already being used in the scientific literature of that time. With Suess' conception of these three new spheres, the Earth could be viewed as comprising four distinct yet mutually overlapping domains. Biosphere is conceived to contain all forms of life that have evolved on the planet Earth. However, as human activity intensified, first through agriculture and forestry, and then industry, accompanied by spectacular innovations in science and technology, the French mathematician and natural philosopher Edouard Le Roy (1870–1954), the French paleontologist, Jesuit and philosopher Pierre Teilhard de Chardin (1881–1955), and the Russian geochemist Vladimir I. Vernadsky (1863–1945) proposed that the biosphere was being transformed into a noösphere, or the sphere of (human) mind. This chapter examines the significance of the biosphere concept that enables us to make a holistic visualization of all life on the Earth (and the ecosystems where they are found). The implications of the ongoing transformation of the biosphere, which represents inclusivity of the biotic and the abiotic, into a noösphere, which effectively became a technosphere dominated by humans, vis-à-vis the concepts of ecocentrism and anthropocentrism, are also discussed.

DOI: 10.4324/9781003481058-8

8.2 Eduard Suess and the Birth of the Biosphere Concept

Eduard Suess is regarded as having made pioneering contributions to pale-ogeography and theories of plate tectonics. He proposed the existence of Gondwanaland, a super-continent that existed about 200 million years ago. It comprised the present-day Antarctica, South America, Africa, India, and Australia, which were joined in a single large landmass. Suess also con-tested the widely held contention at that time that the vertical movements of the Earth's crust caused the formation of mountains. Instead, he argued that it was the horizontal movement of the lithosphere – that is, the Earth's crust – that resulted in the formation of mountains. These ideas formed the basis of the theory of plate tectonics by Wegener in 1912 (Richardson 2022). Suess first proposed the term biosphere in the year 1875 in his book on the formation of mountains – *The Origin of the Alps (Die Enstebung der Alpen)*. He again mentioned this term in 1909 in *The Face of the Earth (Das Antlitz der Erde)*. In his words translated from the original German, Suess stated "One thing seems strange on this celestial body consisting of spheres, namely organic life. But the latter is limited to a determined zone, at the surface of the lithosphere". While observing that organic life is con-fined to a zone on the surface of the lithosphere, Suess also noted the exist-ence of life in the interaction zone of the lithosphere and the atmosphere. The root system of plants penetrates the soil, while the shoot and leaves project upward for respiration. Therefore, Suess concluded that "on the sur-face of the continents we can distinguish a self-maintained biosphere (*eine selbständige Biosphäre)*". He also subscribed to the idea of the "*solidarity of all life*" put forward by Carl Rokitansky, a well-known Viennese patho-logical anatomist at that time. Suess also maintained that the biosphere was "concerned only with life on this planet and all the conditions in regard to temperature, chemical composition and so forth necessary for its existence", not considering the "speculative hypotheses" on the possible presence of life in other celestial bodies. Therefore, according to Suess, "the biosphere is a phenomenon limited not only in space, but also in time" (Samson and Pitt 1999, 15; Suess 1999, 23).

8.3 Vladimir Ivanovich Vernadsky and the Development of the Biosphere Concept

Vladimir I. Vernadsky (1863–1945) was a Russian geochemist and nat-ural philosopher, who is known for fully developing the biosphere con-cept. He is also considered as one of the founders of the subject field of biogeochemistry. Vernadsky contended that the layer that separates the face of the Earth from the atmospheric medium is the biosphere, which forms a film over the Earth's crust and serves a very important role in the planetary functions. The biosphere forms a well-defined "*geological envelope*", which is distinct from the other geological envelopes, that is, the atmosphere, lithosphere, and hydrosphere. It is also the only envelope that is penetrated by "cosmic energy", the major source of which is the sun (Vernadsky 2000–2001, 23).

Thus, the biosphere could be viewed as a dynamic inter-related macrosystem, which includes those portions of atmosphere, hydrosphere, and lithosphere where life exists in constant interaction with the physical environment. It is the sum total of all the ecosystems, landscapes, and biomes. The mind-boggling biodiversity is a defining characteristic of the biosphere. It also represents the zone of transformation, storage, and reallocation of radiant energy. In this way, it "plays an extraordinary planetary role" (Vernadsky 1998, 44). The green part of the living matter photosynthesizes, multiplies, and constantly extends the limits of the biosphere. He further stated that the "whole living world is connected to this green part of life by a direct and unbreakable link". This also makes this "totality of living organisms … a unique system" (Vernadsky 1998, 58). Vernadsky coined the term "living matter", which "is the aggregate of all its living organisms" (Vernadsky 2000–2001, 22), and which forms "a thin but more or less continuous film" on the land surface, and can be found in terrestrial ecosystems such as forests and fields, and the ocean. It is also present under the land surface, approximately up to a depth of around three kilometers. In a biospheric context, Vernadsky viewed life as "living matter", which takes into account the total mass of all living organisms in the biosphere, and not as distinct species and individuals. Besides living matter, the biosphere also has "inert matter", or in other words, inanimate matter. The total mass of inert matter is much higher than that of living matter. However, these two categories of matter do not exist in isolated compartments. There is a continuous movement of different atoms of inert matter to living matter. Similarly, living matter also contributes atoms to inert matter. Consequently, Vernadsky suggested that "biogeochemical phenomena" are "the basis of the biosphere" (Vernadsky 1945, 1, 5–6, 9). Thus, the biosphere concept emphasizes the role played by all living organisms ("living matter") – plants, animals, and microorganisms, large, small, or microscopic – not only in the evolution of the biosphere itself, but in modifying and shaping the atmosphere, the hydrosphere, and the lithosphere, through their concerted interactions with the abiotic components of the biosphere ("inert matter"). Vernadsky also stated that the living organisms comprising the biosphere are the outcome of "complex processes" taking place over a long period of time, "and are an essential part of a harmonious cosmic mechanism" (Vernadsky 1998, 44). Regarding the position of humans in the biosphere, Vernadsky believed that humans are "geologically connected" with the biosphere, which is the "terrestrial envelope where life can exist", and they "cannot be separated from it" (Vernadsky 1945, 4).

8.3.1 Humans and the Biosphere

In the course of his studies on the properties of the biosphere and its evolution, Vernadsky realized that though during the course of the evolution of life on the Earth, the geochemical history of carbon as well as some other elements underwent changes over geological time; these changes were almost exclusively confined to the living matter. The non-living mineral associations were largely conserved during the long evolutionary history of the Earth. However, this picture changed in the present geological epoch, which Vernadsky termed as "the psychozoic era" or the "age of reason". In this epoch, the human species through the process of increasingly intensive agriculture caused "the exchange of atoms of living matter with inert matter". Vernadsky termed this as "the action of the conscious and the collective spirit of humanity on the geochemical processes". Humans have changed the geochemistry of metals, synthesized new compounds, and released many free metals into the environment. Besides metals, other chemical compounds produced due to combustion of coal, metallurgical processes, and others have modified the geochemical cycles, disrupted many natural processes, and added new ones. This led Vernadsky to reason that "With the arrival of man, a new geological force has certainly appeared on the planet's surface" (Vernadsky 1999, 27).

While acknowledging the path-breaking contributions of V.I. Vernadsky toward the shaping of the biosphere concept, Polunin and Grinevald (1999) wrote that the vital role of the biosphere in providing various services that support the existence of both human and nonhuman components of nature is now universally recognized. However, Vernadsky was the first to clearly explain this idea of a unified whole many years ago. With the passage of time and enhancement of human knowledge about the biosphere, the need to protect it from the various potential hazards that range from a nuclear conflict to ozone layer depletion, chemical pollution, and global climate change is becoming more and more evident. The lasting relevance of Vernadsky probably lies in this realization.

8.4 The Transition from the Biosphere to the Noösphere

Humans are also an integral part of the biosphere, and the human species plays its role in maintaining the biospheric functions along with its nonhuman members. This primary role of humans has changed with the advent of the industrial human, which started to modify and remodel not only the atmosphere, hydrosphere, and lithosphere, but also degraded various biospheric habitats, and decimated or even exterminated many nonhuman species occupying these habitats. These anthropogenic phenomena have become so accentuated that the biosphere is believed to have undergone a transition to the noösphere or the sphere of (human) mind, a concept that has been discussed in more detail in the later sections of this chapter. If it is indeed so, then the question arises whether the role of the human species is to become a master of the biosphere and indiscriminately interfere with its structure and functions primarily for its own benefit, or it should take a more benign and participatory stance.

8.4.1 The Noösphere

The special role of the human species is manifest in the concept of the noösphere, which comprises the human mind and intelligence. As mentioned earlier, Edouard Le Roy, Pierre Teilhard de Chardin, and Vladimir I. Vernadsky are jointly credited with the conception and formulation of the noösphere concept. However, ideas advanced by the French philosopher and Nobel laureate Henri Bergson (1859–1941) are thought to have greatly stimulated the development of the noösphere concept by Edouard Le Roy and Teilhard de Chardin. Bergson put forward the idea that all life that evolved on the Earth are attached to matter. Furthermore, the evolution of life marked a constant struggle between consciousness, that is, energy, and matter. Consciousness remains confined to matter or entirely trapped by matter in the case of all other forms of life except in humans, where it breaks free from this confinement. This comprises the uniqueness of the human consciousness and sets humans apart from all other beings. Based on this core idea, Edouard Le Roy hypothesized that it is in the fitness of things to place humans above the other forms of life in a dominant position, though he emphatically stated that this dominance did not imply any "uprooting" of the others. It is, therefore, possible to imagine "a human sphere, the sphere of reflection, of conscious and free invention, of thought in its pure sense: the sphere of the mind or noosphere". This sphere is at a level "higher than the animal biosphere". The concept of noösphere of Le Roy views humans as comprising a "new order of reality" and forming "an ensemble that is also vast, original and new". Le Roy also said that two great events occurred during the Earth's history: one is the "vitalisation of matter" and the other is the "hominisation of life". The two spheres, namely, the biosphere and the noösphere, remain as "two wholes of equal amplitude". The noösphere flows out of the biosphere and covers it in a separate layer. The transition from the biosphere to the noösphere is marked by "the hominisation of life" (Le Roy 1999, 65, 66, 70).

Pierre Teilhard De Chardin also contributed toward the noösphere concept. He had worked in close association with Le Roy and many of their ideas on the issue of situating humans in the ladder of life and in the biosphere as a whole were conceived together. He argued that the evolution of humans that began at the end of the Tertiary period also signaled the appearance of "a new layer, the 'thinking layer'", which has since expanded to form a world "above the world of the plants and animals". This new realm situated above the biosphere is the noösphere. In fact, so striking is the formation of this thinking layer which Teilhard De Chardin imagined that if an extra-terrestrial looked at the Earth, the most characteristic feature that would attract its attention "would be, not the blue of the seas or the green of the forests, but the phosphorescence of thought". He also said that it would be unjust to classify and characterize humans merely as a distinct order or branch in the tree of life. "With hominisation, …. we have the beginning of a new age. The Earth 'gets a new skin'. Better still, it finds its soul" (Teilhard De Chardin 1999a, 71–72).

Along with Edouard Le Roy and Pierre Teilhard de Chardin, Vladimir I. Vernadsky also contributed significantly toward the tenets of the noösphere concept. However, though the three thinkers jointly developed this concept, the perspectives with which they conceived it are marked by certain differences. Both Le Roy and Teilhard de Chardin were religious men, and especially the latter was spiritually inclined. Teilhard de Chardin conceived the noösphere as a "thinking layer", which is placed above the nonhuman, but living, biosphere. Le Roy also held similar views and situated the noösphere on a higher plane. On the other hand, Vernadsky was a materialist who based his concept on a strictly scientific basis. His noösphere and the human species were firmly linked to the biosphere (Samson and Pitt 1999). Vernadsky also wrote that humans cannot be separated from the biosphere, because it provides the materials and energy needed for the survival of all living matter. Thus, there is a geological connection between humans and the "material and energetic structure" of the biosphere (Vernadsky 1945, 4). Despite anchoring humanity firmly to its biospheric base, Vernadsky nevertheless was convinced that the biosphere and humanity were entering a new phase in their history. In a letter to Professor A. Petrunkevitch, he wrote that he was highly optimistic about the "historical" as well as "planetary change" that was already set in motion. He asserted that "We live in a transition to the noösphere" (Vernadsky 1945, 1). In this conviction about the emergence of a noösphere, he held the same belief as that of Le Roy and Teilhard de Chardin. In one of his last articles on the biosphere and the noösphere, which was translated to English and published posthumously, Vernadsky wrote about the phenomenon of "*cephalization*" during organic evolution, a concept put forward by the American mineralogist, geologist, volcanologist, and zoologist James Dwight Dana (1813–1895). This represents the "process of growth and perfection of the central nervous system". It appears to have begun in crustacea (prawns, crabs, and allies), continued in cephalopod mollusks (octopus, squid, etc.), and took place progressively in the vertebrates, culminating in the human species. In terms of the geological role of humans, Vernadsky referred to the Russian geologist and paleontologist A.P. Pavlov, who spoke of an "*anthropogenic era*" that the biosphere was passing through. In this era, humans were "becoming a mighty and ever-growing geological force". Humans have reached every nook and corner of the biosphere, revolutionized communication through inventions such as "radio and television", and made giant leaps in transport with air travel, which has a potential to attain higher speed. All these achievements due to cephalization have resulted in humans becoming *"a single totality in the life of the earth"*. However, Vernadsky was not unaware of the possible pitfalls of the noösphere. He wrote this article under the shadow of World War II, which by that time had engulfed Soviet Russia and eastern Europe, after raging in western Europe and Asia for several years. The horrors of war perhaps led him to say that only if "man .... does not use his brain and his work for self-destruction, an immense future is open before him in the geological history of the biosphere" (Vernadsky 1945, 7–8). Elsewhere, he wrote that humans are "transforming the envelope of the Earth – the geological region of life, the *biosphere*". They have created new biogeochemical processes that were not present earlier. Large amounts of free metals such as aluminum, magnesium, and calcium are being released. "Plant and animal life are being changed and disturbed in the most drastic manner. New species and races are being created. The face of the Earth is changing profoundly". Still, he hoped that the future course of the noösphere "will be grandiose" (Vernadsky 2000–2001, 22–23).

G. Evelyn Hutchinson (1903–1991) was a US zoologist known for his ecological studies on freshwater lakes. He was also an early proponent of the biosphere concept. He ardently supported Vernadsky's concept of a biosphere, and made valuable contributions toward understanding its origin and evolution. Hutchinson was also concerned about human activities leading to premature destruction of the biosphere, which he apprehended was approaching a crisis. The pollutants introduced into the biosphere could greatly hamper its evolutionary progress. Despite these apprehensions, Hutchinson shared the belief of Vernadsky that mankind is undergoing "not only a historical, but a planetary change as well. We live in a transition to the noosphere". Though Vernadsky was optimistic about the capabilities of "the envelope of mind" or the noösphere that "was to supersede the biosphere", Hutchinson, writing a couple of decades after Vernadsky's death, was not so hopeful about the fate of the biosphere after observing the "mindless ... changes wrought by man on the biosphere" (Hutchinson 1970, 53). Around this time, Eugene P. Odum, who is often called the father of modern ecology, also reminded us that though the biosphere is a cybernetic system, its homoeostatic mechanisms cannot protect it from stress beyond its threshold level of tolerance. In such conditions, the system cannot revert to its original state, but has to settle at a higher equilibrium level. Effects of pollution, ozone layer depletion, and the ongoing increase in the concentration of atmospheric carbon dioxide are examples of stress that the biosphere has to experience due to human activities. Odum reasoned that due to the extraordinary development of the central nervous system, humans have acquired the ability to modify the structure and function of ecosystems. Humans have become "a mighty geological agent" which led Vernadsky to suggest that the noösphere (Greek *noos* meaning mind), which is dominated by the mind of man, has now replaced the biosphere, which is a domain of all living organisms. However, Odum warned that "this is dangerous philosophy", which could have detrimental consequences, since it propagates that humans can manage everything and can understand the consequences of all their actions (Odum 1971, 35). E.P. Odum's brother, Howard T. Odum also found the noösphere to be untenable because of the fact that "power flows" created by humans might have replaced those of nature in industrialized areas, but such pockets are able to survive because of the purification by unpolluted areas of the biosphere through various mechanisms (Samson and Pitt 1999, 8). The environmental crisis that has intensified in the recent years due to global climate change bears testimony to the fact that the noösphere has failed to direct the biosphere along a safe route, and that it may be a fallacy to assume that all natural processes could be human-managed, as Odum had prophesied. On the other hand, Hutchinson, who was more optimistic, believed that "Vernadsky's transition in its deepest sense" could ensure a healthy continuity of the biosphere (Hutchinson 1970).

8.5 Noösphere or Technosphere?

Today, more than 100 years after the noösphere concept was put forward by three leading thinkers of their time, one might ask the question whether the noösphere actually got transformed into a realm at a plane higher than the biosphere, with elevated moral and spiritual values, as conceived by Le Roy and Teilhard de Chardin. Or whether mankind as a "mighty geological force" could transform the biosphere into a more habitable place for its human and nonhuman inhabitants (living matter of Vernadsky), better supported by the "inert matter" forming the abiotic base of the ecosystems. Without being overtly critical, we can perhaps say that while a "thinking layer" or an inventing layer has definitely been created, the environmental crisis in terms of toxic contamination and pollution of the environment, large-scale decimation of biodiversity, and climate change accompanied by rise in temperature and sea level, melting of glaciers and ice sheets, and an increased frequency and severity of floods and droughts, have largely belied the expectations of the three thinkers.

We can also perhaps argue that the environmental crisis happened because the concept of noösphere, even in its least human-centered version, can be best termed as weakly anthropocentric. Though it also promises to avoid any 'uprooting', thereby advocating a harmonious existence with nature, the terms of that co-existence are essentially dictated by humans, keeping human interests at the uppermost.

The vision of a noösphere has in reality become the vision of a "technosphere", where technology will dictate almost all actions, including human relationships with other species, and the integrity of natural systems. Such a technologically managed nature – the "second nature" – is totally under human control (Sessions 1995). This technology-oriented thinking is evident in Le Roy's words that the human species can be "defined by instruments detached from the body [which] in reality, is the natural in the artificial". He believed that the full biological reality can only be understood if mankind is not separated "from his tools, his real organs or his technology: his real functions" (Le Roy 1999, 61). Teilhard de Chardin also held similar views. In his *The Appearance of Man*, he noted that technology has made remarkable progress, and become a "*geo-technology*" spread worldwide. Consequently, "Man-as-individual" is centered on "brain", and "Man-as-species (or the noösphere)" is centered on "technology" (Teilhard de Chardin 1966, 236, 241). In *Activation of Energy*, he expressed the idea that "general technology is not merely a sum of commercial enterprises", so as to become a "mechanical dead-weight" to be carried by humans. In his vision, the totality of all technological processes is to be "combined reflectively ... to preserve in men the state of consciousness" needed to reach the goal of "aggregation and conjunction". He further said that "through technology, evolution is making a fresh bound, ... and becoming reflective", and he expressed the conviction that in its relationship with consciousness, technology would enable humans to "develop powers of a grander order – of a spiritual order – and to force us to make up our minds on the question of a religion" (Teilhard de Chardin 1971, 159, 161, 163). Though Teilhard de Chardin kept his faith in the potential of the "thinking layer" of his noösphere, he was aware of the fact that due to an unprecedented increase in both population and consumption, humans may exhaust the reserves of iron, oil, and coal, and create large gaps between the cultivable land area and the demand of a burgeoning population. Nevertheless, he laid his hopes on the development of nuclear energy as a substitute for the conventional forms of energy, and the rapid progress in "organic synthesis". This faith made him assert that "the survival of material plenty ... can only be assured by still more science, and still more ambition, and still more wisdom" (Teilhard de Chardin 1966, 250–252).

As said earlier, E.P. Odum termed the noösphere a dangerous philosophy. Though the three architects of the noösphere concept, namely, Le Roy, Teilhard de Chardin, and V.I. Vernadsky, did not imply that the formation of the noösphere would involve any damage of the biosphere or its nonhuman inhabitants and natural systems, in reality, we find large-scale decimation of the other members of the biosphere and degradation and destruction of entire ecosystems by the action of humans. It is possible that it is an inevitable outcome of viewing humanity as an entity placed above and beyond the jurisdiction of the biosphere's laws. It is also possible that the direction of the evolution of the noösphere was led along a wrong track sometime during its evolution.

Long before the biosphere and noösphere concepts were proposed, world-views upholding human subjugation of other biological species and systems ran strong, especially in the west. Forming the basis of such worldviews was the "man-nature dualism" propagated by many western scientists and think-ers, especially René Descartes. Thus, the biosphere of Suess was viewed more as an anthroposphere, and later, with the advent of more and more powerful technology, as a technosphere in the western countries. These worldviews dominated the societal attitude toward nature, and spread in the colonies as well. Such a worldview runs strong till today among many decision-makers, scientists, and administrators the world over. With colonial expansion and the spread of global trade, technology became increasingly subservient to big business, and this gave rise to more ecological and ethical contradictions that lie at the root of the present environmental crisis. However, since the publication of Rachel Carson's *Silent Spring* in 1962, the "other voices" of Spinoza, Thoreau, Leopold, and others began to be heard again, and alternative roles of man as "steward", "partner", or "participant" in the bio-sphere were put forward by many supporters of ecocentrism or even "soft" anthropocentrism (Zweers 1994).

Despite the worldwide acceptance of biocentric and ecocentric princi-ples by a considerable number of thinkers, the "dream" of having techno-logical solutions to almost all human problems has not been discarded. In recent years, biotechnological solutions to challenges of food scarcity, pest depredations, environmental pollution, and biodiversity decline have been advanced. However, this has also generated an intense and often acrimoni-ous debate on the implications of biotechnological research and application. Broadly analyzing this debate, three distinct viewpoints could perhaps be recognized. The first viewpoint gives a blanket assurance that biotechno-logical research and application should continue without any hindrance as it poses no threat to the ecological, social, cultural, and religious systems. The proponents of the other viewpoint that is diametrically opposite are skepti-cal or even afraid of such research and application, which they advise are not to be pursued at all because of their adverse effects on all or some of the aforesaid systems. The third group that lies somewhere between these two extremes advocates the exercise of caution and adoption of ecological, socio-cultural, ethical, and even legislative safeguards before and while going "full steam" with all technological and biotechnological research and applica-tion. "Meanwhile, technology remains the major, if not sole, yardstick in classifying nations as 'developed', 'developing' and 'underdeveloped', and in identifying communities and regions within a given country as 'advanced' or 'backward', ignoring all other social, cultural and ethical attributes". The level of biotechnological development is already being used as a yardstick for progress and development (Gupta 2008). The new yardstick that is emerging is almost certainly going to be that of the advancements made in artificial intelligence (AI) research, especially artificial general intelligence (AGI), and artificial super intelligence (ASI). Many AI experts, scientists, and industry leaders like Elon Musk have expressed grave concern over the development of AGI/ASI, and called for at least a temporary embargo on their development till an appropriate regulatory framework can be put in place. A recent document by Open AI (Altman et al. 2023), a leading research laboratory and company in the field of artificial intelligence, has called for the urgent framing of rules of governance for superintelligence. ASI can be explained as AGI that could go beyond human control. This risk calls for regulatory measures for its development and governance.

8.6 The Nature of Technology

Opinions vary on whether the small- and large-scale alterations of both biotic and abiotic natural systems wrought by humans through the application of technology have been mostly harmful or beneficial (for humans and/or nonhuman nature). Therefore, the nature of technology needs to be understood from the right perspective in order to set objectives that are more tuned to achieve sustainability and conservation.

The German philosopher Martin Heidegger did not advocate a total rejection of science and technology or consider them as purely harmful to life and society. Heidegger delved deep into the "essence" – that is the very nature – of technology. One definition of technology is that it is a "human activity", while another is that it is "a means to an end". "The manufacture and utilization of equipment, tools, and machines comprise technology". Besides, the end-products of manufacture and the purpose for which these have been manufactured are also a part of technology. However, the definition of technology, "as a means and a human activity, can therefore be called the instrumental and anthropological definition of technology" (Heidegger 1977, 4–5). In this sense, both primitive and modern technology are "means to an end". However, Heidegger said that it is very important to manipulate "technology in the proper manner", and try to have a "spiritual" control over technology. This is a vital task, especially in view of the growing apprehension that technology may go beyond human control. Heidegger said that both natural phenomena, such as the blossoming of a flower, and certain types of technology, such as the creative work of an artisan, essentially comprise the phenomenon of "bringing forth" or "poiesis". However, in contrast to the blossoming of a flower, which is a bringing forth "in itself", the creation of an artisan or an artist represents a "bringing forth not in itself, but in another.... in the craftsman or artist". "Bringing-forth" causes "something concealed ... (to come into) .... unconcealment", and this "coming" is in the realm of "revealing". Therefore, technology is "no mere means" ... but "a way of revealing" (Heidegger 1977, 10–11). With this interpretation of technology, its essence opens up in the "realm of revealing". This means that "It reveals whatever does not bring itself forth". Technology, including modern technology, is also "revealing", but the revealing of the latter "does not unfold into a bringing forth in the sense of poiesis". On the contrary, the bringing forth of modern technology is governed by the principle of "challenging". This approach demands energy and resources from nature to be extracted and stored. The land is mined for coal, oil, and minerals, which are extracted and stockpiled. Modern farming turns simple, traditional agriculture into a practice that challenges the soil to produce crops that sustain the "mechanized food industry". Not only that, it also challenges the soil to alter its natural processes and cycles. Heidegger gave the example of River Rhine, which is dammed up to turn the river into a "water power supplier". This act of challenging by modern technology, which results in unlocking, transforming, storing, distributing, and finally switching on and off the energy of nature, turns nature into a "standing-reserve". The ordering of nature as "standing-reserve" is governed by "enframing". Heidegger explained enframing as "the way of revealing which holds sway in the essence of modern technology". In other words, we might say that enframing is the principle which guides the direction and choices of modern technology, and when people opt for this principle, they treat the natural capital as standing reserve. Heidegger further explained that the essence of technology has "nothing technological" about it, unlike the tools, components, and equipment used in technology, but it is "the challenge of enframing" to which technological activity responds (Heidegger 1977, 12–21). William Lovitt, the English translator of Heidegger's *Questions Concerning Technology*, commented that "Enframing" has the characteristic of treating everything as "supply", leading to the ordering of everything as "standing-reserve", which leaves nothing in its original free state of existence.

Just as art loses its purity and real purpose when it deviates from the path of revealing, turns into a standing reserve, and subjects itself to the dictates of trade, profit, and political expediency, technology also is diverted from its real path of "revelation" to that of "enframing", and in the process is itself enslaved.

8.7 Biosphere and Noösphere in the Anthropocene

Coined by American biologist Eugene Stoermer in the late 1980s, and popularized by Dutch chemist and Nobel laureate Paul Crutzen in 2000, Anthropocene has been officially recognized as a formal geologic epoch after the Holocene epoch by the Anthropocene Working Group (AWG) of the International Union of Geologic Sciences (IUGS) and the International Commission of Stratigraphy (ICS) in 2016. The year 1950 CE has been recommended as the starting point of Anthropocene by the AWG, though various other dates such as the rise of agriculture 10,000–15,000 years ago, the extinction of large mammals such as mammoths 14,000 years ago, and 1784 CE, the year James Watt invented the steam engine, have also been suggested as the point of onset of the Anthropocene (Rafferty 2023). Among the many precursors of the idea of Anthropocene, especially notable are the noösphere concept, and the "Anthropozoic era" of the Italian geologist Antonio Stoppani in 1873 (Crutzen 2002; Hamilton and Grinevald 2015). The question that is posed in this section concerns the direction that the noösphere has taken and the major factors influencing that direction in the Anthropocene. Instead of a "thinking layer" of Teilhard de Chardin that is superimposed on the non-human biosphere, humanity in the form of *Homo faber* (Latin: human being as the maker or creator) or the technological human, led the human progress in the industrial period and in the Anthropocene (Hamilton and Grinevald 2015). At the same time, technology, instead of blossoming as "revelation" or "bringing forth" in a Heideggerian sense, mostly developed as "enframing", which was in turn subservient to trade, profit-making, political, and economic controls. Paul J. Crutzen in his path-breaking one-page publication in *Nature* pointed out that "A daunting task lies ahead for scientists and engineers to guide society toward environmentally sustainable management during the era of the Anthropocene. This will require appropriate human behaviour at all scales, …" (Crutzen 2002). Unfortunately, despite scientists informing world leaders about the imminent threat of climate change from 1970s to 1980s, the latter paid attention much later, and, even when they did, went about it in a half-hearted manner (Chakrabarty 2009). The Paris Agreement was signed in 2015, more than two decades after the UNFCCC agreement in 1992, and yet the outcome is uncertain because it is still not clear as to how honest is the intent of the political leaders and the industry captains. Despite ecologists talking about ecological boundaries and limits of resource utilization, with similar warnings coming from reports such as *Limits to Growth* in 1972, the path of development along that of unbridled expansion has been continued to be followed, both in developed and not-so-developed societies. Chakrabarty (2009) referred to Giovanni Arrighi's 2007 book *Adam Smith in Beijing: Lineages of the Twenty-First Century*, which raised the question of limits to capitalism. However, the responses to such questioning have tended to remain lukewarm. The seminal document on environmentally sustainable development: *Our Common Future* by the World Commission on Environment and Development (WCED) in 1987, also called the Brundtland Report, proclaimed that "What is needed now is a new era of economic growth – growth that is forceful and at the same time socially and environmentally sustainable". While acknowledging the negative impacts of energy-intensive technologies, pollution, and environmental degradation, it nevertheless declared: "We see instead the possibility for a new era of economic growth, one that must be based on policies that sustain and expand the environmental resource base" (United Nations 1987). The implications of such a policy are that while implementing, the executors always lay the first priority on growth, which is rarely compromised, while factors such as prudence in resource use, environmental protection and restoration, and sustainability are compromised at the slightest pretext. Political expediency comes into play, citing the "pressing need" of a burgeoning population, and their aspirations for a more prosperous living. The fact remains that so long the attitude remains anthropocentric, the nonhuman environment including its biodiversity is bound to receive a stepmotherly treatment from humans. The emphasis on growth in the WCED report is beset with ambiguity, because on the one hand it recognizes the dangers of excessive natural capital utilization and stresses on sustainability, while on the other hand, it calls for a five- to ten-fold expansion in economy to improve the living standards of the poor. It is obvious that a distinction has to be made between development and growth, where development signifies qualitative improvement in contrast to growth, which stands for quantitative increase. The Brundtland Report justifiably stressed "producing more with less", which hinges on resource conservation, resource use efficiency, improvements in technology, and recycling. It also advocated control of population growth, and "redistribution from overconsumers to the poor". However, it is rather ambiguous on the need to stop any further physical expansion of the economy, and aim for qualitative improvements within the same scale. Political expediency is perhaps at play for this ambiguity in the Report (Costanza et al. 1997, 24, 112).

8.8 "Economy of Permanence" versus Economy of Violence

More than six decades ago, J.C. Kumarappa, an associate of M.K. Gandhi, called for replacing the "economy of violence" that was prevalent then (and largely continuing till today), with an "economy of permanence" (Kumarappa 1957). Kumarappa spoke about renewable and nonrenewable resources, calling them as "permanent" and "transient", respectively. Nature's permanence in living beings is illustrated by the "cycle of life" which is characterized by the cooperative interaction of many factors that enable life to perpetuate. Plant seeds germinate to give rise to new plants, leaves that are shed, and other litter decompose to fertilize the soil, and the young plant grows to again produce seed of its own, and unless this cycle is broken, life continues and is permanent in this sense. Kumarappa also noted that many types of life forms interact with the inanimate matter and among themselves, which he termed as "vital cooperation existing between soil, plant and animal life". This sort of cooperation is evident in the role of earthworms in making the soil fertile, and the bees and butterflies in pollination. In exchange, they receive their nutrition as a kind of "honest wages" paid by nature. Birds and other animals help in the propagation of plants by consuming fruits and then passing the seeds, sometimes far away. He, therefore, surmised that

> nature enlists and ensures the cooperation of all its units – the mobile helping the immobile, and the sentient the insentient. Thus all nature is dovetailed together in a common cause. Nothing exists for itself. When this works out harmoniously and violence does not break the chain, we have an economy of permanence.
>
> *(Kumarappa 1957, 9–10)*

Interpreted in this sense, human interference with nature in terms of mindless alteration and destruction of ecosystems, and decimation of the populations of nonhuman species, by treating them as mere resources for profit and growth, can be dubbed as an economy of violence, and can never culminate in sustainability. In the context of the present model of development vis-à-vis the issue of sustainability, the three basic criteria for maintaining natural capital and environmental sustainability, which were identified by the ecological economist Herman Daly, may be examined (Daly 1990, cited in Costanza et al. 1997).

1. In the case of renewable resources, the rate of harvest should not exceed the rate of regeneration (sustainable yield);
2. The rates of waste generation from projects should not exceed the natural assimilative capacity of the environment (sustainable waste disposal); and
3. In the case of nonrenewable resources, their depletion should be compensated by developing comparable amounts of renewable substitutes for a given resource.

If an appraisal is done of human performance with respect to the first criterion, that is balancing harvesting with regeneration, UN FAO data reveal that ~ten million hectares of forest are cut down every year, with a net loss (difference between forest loss and forest regeneration) of 4.7 million hectares. While some countries such as China and India are increasing their forest areas, huge tracts of tropical forest in South America, tropical Africa, and Myanmar, Thailand, and Indonesia in Asia are lost every year (Ritchie and Roser 2021). With regard to the second criterion of waste generation, about 353 million tons of plastic waste were generated worldwide in 2019.

Of this, 55 million tons were collected for recycling, though only 33 million tons were ultimately recycled (Alves 2023). Thus, a large volume of plastic waste is not managed properly, and these find their way into different water bodies and the sea. They cause entanglement, injury to gastrointestinal tract, and toxic effects to marine organisms, affecting and even killing thousands of marine mammals such as whales, dolphins, and porpoises annually. Similarly, toxic metal wastes are contaminating soil, water, and air, and imparting toxic effects on organisms, including humans. All the plastic, metal, and other hazardous wastes are byproducts of human economic activities. Looking at these issues from the point of view of Kumarappa, an economy that destroys natural systems like forests, produces wastes that it cannot handle adequately, and inflicts irreparable damages to plants, animals, humans, and entire ecosystems like the ocean may be termed as an "economy of violence". Rutherford et al. (2007) spoke about "an overarching definition of violence", which according to the World Health Organization (WHO) is

> the intentional use of physical force or power, threatened or actual, against oneself, another person, or against a group or community, that either results in or has a high likelihood of resulting in injury, death, psychological harm, maldevelopment or deprivation.

Thinking ecocentrically, one can extend this definition to include nonhuman organisms, communities, and ecosystems to give an inclusive, ecological definition of violence produced by a "violent" economy and culture.

8.9 Is it Possible to Have an Ecocentric Model of Development?

The mode of functioning of the economic and political systems affects the integrity of natural systems. Governments of the world have not exhibited enough political will to cut down on greenhouse gas emissions, or degradation of natural ecosystems like forests to prevent loss of biodiversity. This lack of political will often arises from economic compulsions, which compel the governments to meet the material demands and aspirations of the human populations they serve by any means. Sustainable development conceptually imposes a limit on the extent of extraction and utilization of various components of nonhuman nature, which are, however, expendable, and can give way to other, more pressing priorities. This has been possible because nature has been stripped of all its values to be reduced to natural resources that only possess economic and political values. The Brundtland Report acknowledged the limits of exploitation, when it recognized that "nature is bountiful, but it is also fragile and finely balanced". It further stated that humans are crossing many thresholds that "cannot be crossed without endangering the basic integrity of the system". The Report, however, then came to a compromise by contending that the limits are not absolute but "imposed by the [prevalent] state of technology and social organization on environmental resources". Since both technology and social organization can be "managed and improved", it is possible to have "a new era of economic growth" (United Nations 1987, 15, 29). In reality, the limits imposed by the national governments are highly influenced by the societal demands and political compulsions. These factors make the transition from an anthropocentric-technocentric interpretation of sustainable development ("weak sustainability") to an ecocentric one ("strong sustainability") beset with many hurdles (Montani 2007).

Hurdles can also be posed by some of the dominant approaches in economics. Philosopher Adam Dickerson, while examining whether there could be ecocentric paths toward economic policies, cited Léon Walras (183–1910), the French mathematician-economist who is regarded as the founder of the modern theory of general economic equilibrium. Walras believed that "minerals, plants and animals are things" because they are neither conscious of nor master of themselves. "Whatever is conscious of itself and master of itself is a person". "Man alone is a person". He further reasoned that rationally speaking, "the purpose of things is under the dominion of the purpose of persons". Hence, humans are the masters of all things (Walras 1954, cited in Dickerson 2020). In the backdrop of such assertions, Dickerson raised the question whether anthropocentrism is built into economics, or whether an "*ecocentric* economics" is at all possible. He believed that the concept of "human mastery over nonhuman nature" is not necessarily indispensable to economics. However, the practice of attaching a market price tag to all processes and goods, such as the "shadow pricing" of non-market entities like natural processes and wildlife is essentially anthropocentric. This practice enables economics (and economists) have the final say in policy debates, but also ends up by valuing nature as a market commodity. When this happens, it lends an anthropocentric bias to economics, because market prices are based on human preferences and valuation. Can economics come out from its anthropocentric confines and turn toward ecocentrism? Dickerson held the view that this transformation would require an attitudinal change "in those social practices and institutions that gives orthodox economics such power and influence" (Dickerson 2020, 5, 8).

An anthropocentric approach in addressing environmental issues is evident in most international programs and action plans for protecting the environment. The United Nations Conference on the Human Environment held at Stockholm during June 5–16, 1972, which can be said to mark the beginning of international action on emerging environmental issues, adopted an anthropocentric approach. Its declaration stressed the "need for a common outlook and for common principles ... in the preservation and enhancement of the human environment". It gave a central position to humans by proclaiming that "Man is both creature and moulder of his environment ... [and] the protection and improvement of the human environment is a major issue which affects the well-being of peoples and economic development throughout". Though it states that pollution and other environmental disturbances are affecting "the ecological balance of the biosphere", the major effects comprised "destruction and depletion of irreplaceable resources; and gross deficiencies, harmful to the physical, mental and social health of man, in the man-made environment, particularly in the living and working environment". Thus, the accent was clearly on human safety and welfare, and if the biosphere was mentioned, it was in the context of its "irreplaceable resources" that were becoming scarce due to human activities, and posed a threat to human societies. Nevertheless, it tried to put a leash on anthropogenic activities by assigning humans the role of steward of the environment. In *Principle 4*, it stated that "Man has a special responsibility to safeguard and wisely manage the heritage of wildlife and its habitat", and "Nature conservation, including wildlife, must therefore receive importance in planning for economic development" (United Nations 1973, 3, 4).

The anthropocentric emphasis of another important international document, namely, the World Conservation Strategy (WCS), is amply illustrated by its "three main objectives of living resource conservation". The first objective is "to maintain essential ecological processes and life support systems". For example, it is essential to protect the soil and help in its regeneration, maintain nutrient recycling, and improve water quality. These measures can ensure "human survival and development". The second is "to preserve genetic diversity.. This would ensure efficient functioning of many vital "processes and life support systems", which in turn are essential for developing breeding programs for the production of improved varieties of crops and domestic animals, advancements of biomedical research and biotechnological innovations, and the flourishing of bio-based industries. Thirdly, the Strategy intended "to ensure the sustainable utilization of species and ecosystems" that support numerous "rural communities as well as major industries". "Fish and other wildlife, forests and grazing lands" are examples of such vital resources (IUCN 1980, VI). Thus, the priority for conservation was not nature *per se*, but those parts of nature which are essential for humans one way or another.

In contrary to these documents, the World Charter for Nature stands out for its biocentric/ecocentric views. This Charter was approved in 1982 by the General Assembly of the United Nations in its 37th Session. It says that: "Mankind is a part of nature and life depends on the uninterrupted functioning of natural systems which ensure the supply of energy and nutrients". The Charter also affirmed that: "Every form of life is unique, warranting respect regardless of its worth to man, and, to accord other organisms such recognition, man must be guided by a moral code of action". Further, the General Principles of this Charter in its *Principle 1* stated that "Nature shall be respected and its essential processes shall not be impaired" (United Nations 1982, 17).

The respect for and recognition of intrinsic value in nature expressed in the World Charter for Nature did not grow further and develop into a trend or paradigm. As said earlier, the Brundtland Report maintained ambiguity in its stand on the intrinsic value of nature, and lobbied for more growth aided by improved technology. Another path-setting document, the "Report of the United Nations Conference on Environment and Development (UNCED)", also known as the Rio Summit – held at Rio de Janeiro during June 3–14, 1992 – was adopted in the UN General Assembly on August 12, 1992. Though this report recognized "the integral and interdependent nature of the Earth, our home" in its preamble, it proclaimed in its *Principle 1* that "Human beings are at the centre of concerns for sustainable development. They are entitled to a healthy and productive life in harmony with nature". Notwithstanding this human focus, the document did introduce several progressive ideals such as more equitable distribution of wealth and resources among the rich and poor nations; wider public participation in decision-making on environmental issues; the sovereign rights and responsibilities of states in matters pertaining to the environment; introduction of the "polluter pays" principle; recognition of the "vital role" of women and "indigenous people and their communities"; and ensuring their participation, because their traditional knowledge systems and practices are invaluable "in environmental management and development". It also asserts that "peace, development and environmental protection are interdependent and indivisible". Another important principle (*Principle 7*) laid down in this document pertained to "the common but differentiated responsibilities" of the States. The onus of ensuring sustainable development all over the world was on the developed countries because of "the pressures their societies place on the global environment and of the technologies and financial resources they command" (United Nations. 1992a, 1–2). The accompanying Convention on Biological Diversity (United Nations 1992b) affirmed in its *Preamble* that it is "Conscious of the intrinsic value of biodiversity" in its very opening statement. It neither elaborated this recognition any further, nor reiterated its commitment toward promoting and facilitating the growth of ecocentric ideals among citizens. The Convention also gave due recognition to "the close and traditional dependence of many indigenous and local communities embodying traditional lifestyles on biological resources", and "the vital role that women play in the conservation and sustainable use of biological diversity". However, it was silent on the ecocentric belief systems and recognition of intrinsic value in biodiversity and natural systems so characteristic of most indigenous communities worldwide.

Twenty years after the 1992 Earth Summit (UNCED), the Rio + 20 United Nations Conference on Sustainable Development was held at Rio de Janeiro, during June 20–22, 2012, where "the Heads of State and Government and high-level representatives" reiterated their "commitment to sustainable development". It aimed to replace "unsustainable patterns of consumption and production" with sustainable alternatives, and had major commitments toward achieving the "Millennium Development Goals by 2015". Its major focus was on human rights and reduction of poverty and human sufferings, and it was not explicit on animal rights or the broader rights of all nonhuman living organisms and natural ecosystems, and their moral-ethical position vis-à-vis that of humans. However, in points 39–41, it recognized "that the planet Earth and its ecosystems are our home", and observing that the rights of nature were recognized in some countries, stressed the need for "holistic and integrated approaches to sustainable development", which could help humans to "live in harmony with nature". The document further accepted that the great natural and cultural diversity of the world "can contribute to sustainable development" (United Nations 2012, 1–4). Washington et al. (2017) conjectured that the recognition of ecocentric principles in point 39 of this document was a fallout of the incorporation of rights of nature in its constitution by Ecuador in 2008 (Republic of Ecuador 2011), followed by the passing of the Law of the Rights of Mother Earth by Bolivia in 2010. Washington and his coauthors also argued that ecocentrism was essential for achieving sustainability, because human societies had been following ecocentric beliefs and principles since a very early stage in their evolution. Ecocentrism makes us ethically richer through an expansion of the reach of our moral community by including nonhuman organisms and life-supporting ecosystems, and by nurturing empathy in humans. Furthermore, ecocentrism is also evolutionarily justifiable as it reflects the links of the human species with the other life forms; spiritually enriching as it is an integral part of many religions all over the world; and ecologically compatible because it reflects the interconnectedness of life and natural ecosystems, as has been reflected in ecological principles (Washington et al. 2017).

8.10 Conclusions

Ecocentric beliefs, rituals, and practices have been the hallmarks of most indigenous societies the world over, and enabled them to live in a sustainable manner in their respective environments (see Chapters 5 and 6 for details). Of course, their lifestyles cannot be judged as "affluent" in today's terms, yet they can fulfill most of their basic needs. Needless to say, such traditional models of development may not be feasible or even practical to adopt in a world that has undergone drastic changes in terms of an increased population load, and vastly improved technologies and knowledge systems that have made possible the lifting of the "lifestyle threshold" higher and higher, from which it would be extremely difficult, well-nigh impossible, for people to revert to the earlier state. However, several recent developments indicate that ecocentric principles are being increasingly accepted at governmental and institutional levels. The examples of the governments of Ecuador and Bolivia, and a shift of emphasis in the 2015 Rio + 20 conference, have already been cited in Section 8.9. Recognition of ecocentric ideals has spread further in the recent years, finding their place in the legislative framework of several other countries and in international institutions. Indigenous laws on resource use are finding increasing acceptability. Some major examples include the recognition of the rights of nature in the national constitution and in laws of several countries such as Ecuador, Bolivia, and Panama; in the legal framework of some indigenous nations as well as counties in the USA; incorporation of the rights of nature in the policies of the Green Party of England and Wales; recognition of the rights of forests and rivers in New Zealand, Colombia, and India; and adoption of a resolution on "harmony with nature" by the United Nations General Assembly on December 14, 2022 (United nations 2023). These case studies and the issues pertaining to the incorporation of ecocentric precepts and principles in local, national, and international policies and governance frameworks are discussed in detail in Chapter 9.

9

ECOCENTRISM IN ENVIRONMENTAL POLICY, LAW AND GOVERNANCE

9.1 Introduction

The preceding chapters suggest that a reorientation from the present anthropocentric toward a more ecocentric approach in most environmental issues – such as conservation, sustainable development initiatives, adaptation to and mitigation of the effects of climate change, prevention and control of environmental pollution, and biodiversity loss – is necessary for a healthy continuation or even the survival of the biosphere. If this change has to take place, then it is essential for ecocentric philosophies to be reflected in the legal and policy frameworks and governance systems. However, if we take a look at the biodiversity and natural habitat conservation policies of countries all over the world, we find that these are traditionally with an emphasis on the use value of biodiversity and natural habitats, with ecosystem services provided to humans largely driving the initiatives for biodiversity and nature conservation. For example, the preservation of genetic diversity is primarily for conserving genes of organisms having (or likely to be having in the future) potential for augmenting food production, yielding medicinal compounds, biofuel or biofertilizer, or with high tourism prospects. Even when a keystone species is prioritized, some human benefit goals are either explicit or hidden. Is this trend changing in the recent times? The answer to this question is sought here by examining the issue of legal rights given to nature and natural entities in different countries in the recent years, because such changes could possibly herald the coming of a paradigm shift toward more ecocentrically oriented policies and governance systems.

DOI: 10.4324/9781003481058-9

9.2 Earth Jurisprudence

While the incorporation of ecocentric principles in biodiversity conservation and other environmental policies is at a nascent stage, positive steps have been initiated in jurisprudence, where Earth jurisprudence has incorporated the recognition of intrinsic value of natural entities and applied them in legislations on various environmental issues. Peter Burdon – an Australian scholar studying Earth jurisprudence – observed that guided by a predominantly anthropocentric Western jurisprudence, our legal systems recognize relationships among human individuals, communities, nations, and other groupings, but rarely extends it to include nature or even nonhuman animals. This definition places humans outside the laws of nature and makes it legal to extract resources such as fish, water, etc., beyond sustainable levels, pursue intensive agriculture in lands not suitable for such activities, or give license to industries to emit large amounts of carbon and other pollutants into the atmosphere. These actions are backed by the contention that nature is "human property". The liberal political theory which attaches great importance to freedom and individual rights over other commitments also influences public opinion in favor of attaching a lower priority to nonhuman nature, compared to the rights of an individual human. Consequently, the nonhuman environment is assessed only on the basis of its human use value, and is not vested with any rights of its own. It was Thomas Berry who first pointed out the inherent flaws of this theory of jurisprudence (Burdon 2011). Thomas Berry (1914–2009) was a Catholic priest, cultural historian, and a scholar of world religions. He also called himself a geologian because of his studies in Earth history and evolution. Deep Ecologist George Sessions described him as an "ecotheologian" (Sessions 1995). He is also regarded as the founder of Earth Jurisprudence, an emerging theory of law (Burdon 2011). In his major work published in 1999, Berry reasoned that because legal rights were exclusively vested in humans, all nonhuman entities enjoyed no rights whatsoever. This lack of rights made them "totally vulnerable to exploitation by the human", and subjected them to "a devastating assault ... by the human" (Berry 1999, 4, 72). This injustice led Berry to plead for a new legal system or a "new jurisprudence". Berry said that

> Within this context the various components of the earth – the land, the water, the air, and the complex of life systems – would each be a commons. Together they would constitute the integral expression of the Great Commons of the planet Earth to be shared in proportion to need among all members of the Earth community.
>
> *(Berry 1999, 61)*

Thus, Earth jurisprudence has emerged as a "legal philosophy", which is a critique of the "western law and jurisprudence [that] reflects an anthropocentric worldview". In this sense, it is "consistent with the widely accepted critique advanced in environmental philosophy" (Berry 1999, 2).

The principles of ecology tell us that humans are not to be considered the center of creation, but a part of the Earth community that includes both humans and nonhumans. Earth jurisprudence develops a legal system that reflects this changed perception by advancing the concept that both humans and nonhuman nature are relevant in a legal system. Evolving around these principles, Earth jurisprudence puts forward a hierarchical system of two broad types of laws. At a higher level is the *Great Law*, which incorporates natural "rules or principles", below which is placed the *Human law*, which is framed by humans and which now needs to be "consistent with the Great law," thus taking into account the structure and functional processes of nature. Any human law, which goes contrary to the Great law and violates natural rules and principles, is "not morally binding on a population" (Burdon 2011). Berry had also argued earlier that law in its present form lays great emphasis on individual human rights that considers the natural world to be meant for human use, and does not recognize legal rights of nonhuman organisms. Such a notion is not acceptable because "The earth belongs to itself and to all the component members of the community. The entire Earth is a gorgeous celebration of existence in all its forms". "A new legal system", therefore, needs to evolve that takes into account "the integral functioning of the earth process, with special reference to a mutually enhancing human-earth relationship" (Berry 1995, 12–13).

The legal systems of humans have been traditionally sanctioning damage to and destruction of natural ecosystems, such as cutting down mountains to mine coal and other minerals; drilling the Earth to extract oil and natural gas; and clear-cutting forests in the Amazon and elsewhere in the world. The root of most of these activities lies in profit-making, often in the garb of development. The climate crisis is a direct fallout of such mindless exploitations. The growing realization that in order to prevent and mitigate the multiple effects of environmental degradation humans need to stop thinking of nature as their property and respect the laws of nature is also permeating the legal system. This is reflected in several court rulings delivered all over the world that recognize the Rights of Nature (RoN) in law. These are indicative of the first signs of a paradigm shift in human-nature relationship (Biggs et al. 2017).

Rights of Nature in law rejects the notion that nature is human property and views humans as just another member of the natural or Earth community that is a part of the "interconnected web of life on earth". It operates on the following premises:

1. It recognizes the right of all living beings "to exist, thrive and evolve".
2. It values and protects nature for its intrinsic worth and not merely for its use value.

Based on these premises, Rights of Nature laws empower nature "to defend and enforce its own rights"; to empower people to defend and enforce (on behalf of nature) its rights; and make it the responsibility of the governments "to implement, defend and enforce the Rights of Nature". Rights of Nature laws differ from the conventional environmental laws in the basic principle that the latter frames rules to minimize the impact of human interferences on natural systems and their flora and fauna, but attach priority to human developmental needs, and are therefore, essentially anthropocentric. Nature does not enjoy any independent rights under these laws. On the other hand, Rights of Nature laws confer rights to nature and enable humans to speak on behalf of nonhuman nature in order to protect it from damage and destruction, and enable it to "exist and flourish". Hence, Rights of Nature laws can be regarded as biocentric or ecocentric in their ethical orientation (AELA 2023).

9.3 Legal Rights of the Earth and Nature in Different Countries and Communities

The concept of Earth jurisprudence has been gaining ground since its inception in 2001. Several legislations based on biocentric and ecocentric ethics have been framed in different parts of the world in the recent years. These include constitutional reforms and revisions, recognition of the rights of Mother Earth, and granting legal status of personhood to natural ecosystems like forests and rivers.

9.3.1 Constitutional Reforms

Ecuador – a small country in South America – was the first country in the world to legally recognize the rights of nature in its national constitution. Chapter Seven of the Ecuador Constitution deals with the "Rights of Nature". In its *Article 71*, it states that "Nature, or *Pacha Mama*, where life is reproduced and occurs, has the right to integral respect for its existence and for the maintenance and regeneration of its life cycles, structure, functions and evolutionary processes". It also empowers "all persons, communities, peoples and nations" to "call upon public authorities to enforce the rights of nature ... The State shall give incentives ... to protect nature and to promote respect for all the elements comprising an ecosystem". *Article 72* declares that "Nature has the right to be restored". *Article 73* affirms that "The State shall apply preventive and restrictive measures on activities that might lead to the extinction of species, the destruction of ecosystems and the permanent alteration of natural cycles". Somewhat contrary to these Articles, which are ecocentric in principle, *Article 74* introduces an anthropocentric condition that "Persons, communities, peoples, and nations shall have the right to benefit from the environment and the natural wealth enabling them to enjoy the good way of living" (Republic of Ecuador. 2011). Whether the insertion of this provision would lead to environmentally unsustainable activities in the garb of meeting the needs of the people will depend on how this Article is interpreted, though such loose and subjective options could always be taken advantage of by the profit-making lobby at the expense of nature. Thus, this could be a loophole in the Constitution that could greatly reduce its efficacy.

9.3.2 Laws on the Rights of Mother Earth

Closely on the heels of Ecuador, which gave constitutional sanctity to rights of nature, another South American nation Bolivia, which calls itself the Plurinacional (multinational) State of Bolivia, enacted the Law of the Rights of Mother Earth in December 2010. This Act was promulgated "to recognize the rights of mother earth", which have to be respected by the state and the society (*Article 1*). The major principles of this Act include harmony of human activities with natural "cycles and processes"; collective good within the framework of the rights of Mother Earth; guarantee of the regeneration of Mother Earth through repair and restoration of any damage inflicted on her; respect and defense of the rights of Mother Earth; no commercialism, which implies that ecosystems and their various functional pathways should not be "commercialized" or made anyone's "private property"; and multi-culturalism that recognizes and respects the cultural diversity of the world, which "seek to live in harmony with nature" (*Article 2*, various clauses). Mother Earth is held sacred by indigenous communities (*Article 3*); and Mother Earth has legal rights that entitles her along with all her components including humans "to all the inherent rights recognized in this law" (*Article 5*). Mother Earth enjoys the rights: i) to life; ii) to the diversity of life; iii) to water; iv) to clean air; v) to equilibrium; vi) to restoration; and vii) to pollution-free living. Both the State and the people have a number of duties and liabilities to Mother Earth that they require to fulfill, and to defend the rights of Mother Earth (Articles 7–10) (Plurinational State of Bolivia. 2010).

This Act was followed by a more elaborate "Framework Law of Mother Earth and Integral Development to Live Well (Ley Marco de la Madre Tierra Y Desarrollo Integral Para Vivir Bien)" in 2012. In *Article 5*, this Law defines Mother Earth as:

> the dynamic living system made up of the community indivisible from all life systems and living beings, which are interrelated, interdependent and complimentary, and they share a common destiny. Mother Earth is considered sacred; it feeds all and is the home that contains, sustains and reproduces all living beings, ecosystems, the biodiversity, organic societies and the individuals that make these up.

The purpose of this Law, as given in *Article 1*, is to establish the vision of integral development for regeneration and restoration of the components and life systems of Mother Earth. The Law also intends to recover and strengthen local and ancestral knowledge. Some of the principles of this Law include: i) rights of Mother Earth as a collective subject of public interest; ii) rights of indigenous nations, native peasants, and intercultural and Afro-Bolivian communities; iii) ensuring Bolivian people to live well with integral development; and iv) non-commercialization of the environmental functions and processes of the components and systems, which are the gifts of the sacred Mother Earth. It also follows the Precautionary Principle; promotes a harmonic, dynamic, adaptive, and balanced approach between the needs of the people and the regenerative capacity of the components and life systems of Mother Earth; and upholds the principles of social and climate justice. It aims to establish non-polluting production processes without toxic effluents; increased use of cleaner and renewable energy; sustainable use of land, forests, agriculture, fishery, and livestock; democratize access to the means of production; conservation of biological and cultural diversity; regulation of mining and hydrocarbon production; wise use and conservation of water; ensure cleaner air and high environmental quality; implement efficient waste management; frame appropriate climate policies and effective adaptation and mitigation mechanisms; and promote intra- and inter-cultural knowledge. The State and the people must guarantee the rights of Mother Earth (Plurinational State of Bolivia. 2012).

Though the world community and the United Nations welcomed the framing of these path-breaking laws, many scholars have been critical of their real purpose and achievements in the Bolivian context. Paola V. Calzadilla and Louis J. Kotzé looked at both the positive aspects of the two laws as well as their limitations and contradictions. They observed that ecocentric concepts and philosophies have rarely found their presence in legal frameworks. The Ecuadorian constitution is perhaps the first constitution in the world to confer rights to nature. Coming to the two Bolivian laws, recognition of ecocentric principles could only be found in the local/county/provincial legislation in some countries such as Canada and the USA, but no national level legislation was in existence. In that sense, the two Mother Earth laws of Bolivia are welcome steps toward an increased enshrining of ecocentric principles in law. The authors expressed the optimism that these two laws can form an initial step toward evolving alternatives to the prevalent anthropocentric orientation in law that sanctions license to environmental destruction in the name of development. These two laws have the potential to generate "political, legislative and academic debates" on the pathways toward a more ecocentric model of development. Despite their positive aspects, these laws also contain provisions that are contrary to ecocentric principles, and are clearly anthropocentric. For example, the first two principles in the 2010 document appear to be somewhat contradictory. The first principle underscores harmony with nature, while the second says that the interest of society should prevail in all human activities, though it adds a rider that such interests should remain within the framework of the rights of Mother Earth. Calzadilla and Kotzé interprets this as an act of prioritizing human interests, which goes against the primary objective of this law to protect the rights of Mother Earth on a priority basis. Furthermore, the principle of no commercialism of living systems and processes is being flouted with unregulated mining and gas extraction, as also stated by Muñoz (2023). It can also be seen that though stipulated in *Article 10* of the Rights of Mother Earth law (2010), an Office of Mother Earth to watch over the enforcement of the rights of Mother Earth has not been created till recently (Calzadilla and Kotzé 2018; Muñoz 2023). Further, the Bolivian Government has neither framed new laws to ensure the rights of Mother Earth as laid down in the 2010 and 2012 laws, nor has it abrogated any law/s, the provisions of which are contrary to the rights and interests of Mother Earth (Calzadilla and Kotzé 2018).

Lorna Muñoz – a Harvard researcher – is also highly critical of the 2010 Mother Earth Law and the 2012 Mother Earth Framework Law. She finds these two laws ineffective in protecting the biodiversity-rich ecosystems of Bolivia. Though the new Bolivian Constitution gives due importance to environmental protection, it also has some anthropocentric principles such as placing the highest value in humans, and giving topmost priority to human interests such as rampant exploitation of natural resources. This has been done by relaxing the regulations for mining and other industries. The impact of over-exploitation of water for mining and agriculture is observed at Lake Poopo, a high-altitude saline lake which completely disappeared in 2015 (Muñoz 2023). Adam Martin, however, points out that despite being dubbed as the poorest country of South America, Bolivia had enacted the "Mother Earth Law" in 2010. This law has drawn mixed responses: while environmentalists have welcomed it, the economists have expressed the fear that it will further strain the economy by posing more challenges. This law is based on the indigenous view of regarding the Earth as a living entity and as mother, who obviously deserves moral rights. However, he also observed that the goals of the Law have been defeated by actions such as the 2020 approval for clearing millions of hectares of forested land for raising livestock and farming (Martin 2023).

Despite these shortcomings in implementation, the passage of the rights of Mother Earth and the Framework laws by the Government of Bolivia along with the constitutional recognition of the rights of nature or *Pacha Mama* by Ecuador in 2008 have had several positive fallouts. For example, the outcome document of the Rio + 20 Conference on Sustainable Development of 2012 declared its recognition of the planet Earth as our home and acknowledged that some countries recognize the rights of Mother Earth, which is a common expression in many cultures all over the world. The Rio + 20 document owes its ecocentric orientation to the legislations in Ecuador and Bolivia. These laws were followed by the recognition of natural ecosystems like forests and rivers as living entities in several countries. Hence, it may be said that though the enactment of the Bolivian laws of the rights of Mother Earth were not accompanied by implementation measures commensurate with their objectives, they could initiate a new paradigm toward ecocentric orientations in environmental legislation worldwide (Calzadilla and Kotzé 2018).

9.4 Recognition of Legal Personhood of Natural Ecosystems

Besides constitutional recognition of nature and the rights of Mother Earth, the other examples of ecocentric legislations comprise recognizing the rights of natural ecosystems such as forests and rivers similar to those given to a human individual.

9.4.1 Te Urewera and Te Awa Tupua Acts, New Zealand

In 2014 and 2017, New Zealand passed two historic legislations in succession. The 2014 legislation pertains to *Te Urewera*, which was a National Park established in 1954 near the east coast of the North Island of New Zealand. Its National Park status was revoked in 2014 in view of the enactment of the Te Urewera Act 2014. It is the home of the Tūhoe people, who settled the tribe's claim over Te Urewera with the Crown in 2013. This was followed by the passing of the Te Urewera Act in 2014, under the provisions of which it ceased to be a National Park, and became Te Urewera – "a legal entity, ... [with] all the rights, powers, duties, and liabilities of a legal person". Through this historic legislation, Te Urewera became the first natural ecosystem in the world to be granted personhood.

In the background of this Act, which is called the "Te Urewera Act 2014", Te Urewera is described as "a place of spiritual value, with its own *mana* and *mauri*" [Mana in Māori culture refers to status and power, while mauri is the life spark or the vital essence of life in all living things. It is passed down from *whakapapa*, which is the line of descent from one's ancestors]. The Act states that "Te Urewera has an identity in and of itself, inspiring people to commit to its care". It has a special significance in the history of the Tūhoe people, to whom it is the heart of the great fish of Maui or the North Island. According to a Māori folklore/creation myth, Māui, a young boy with magical powers, caught a giant fish, chunks from whose body were carved out by his greedy brothers. In this way, they created huge ravines/gullies and mountains in the body of the giant fish, which came to be known as the North Island of *Aotearoa* (Māori name of New Zealand). For the Tūhoe, "Te Urewera is the *ewe whenua*, their place of origin and return, ... their homeland", and is of immense relevance to their "culture, language, customs, and identity". Respecting these beliefs, Te Urewera was declared as "a legal entity, and has all the rights, powers, duties, and liabilities of a legal person". The Act vests these "rights, powers, and duties" to be "exercised and performed on behalf of, and in the name of, Te Urewera by Te Urewera Board in the manner provided for in this Act". The liabilities are also the responsibility of this board (*Subpart 3, Article 11*). Following the promulgation of this Act, Te Urewera loses its former status as a crown land, a National Park, a Conservation Area, and a Reserve. The Te Urewera Board is expected to be guided by Tūhoe concepts of management such as: i) *mana me mauri*, which recognizes the living and spiritual vibes of a place; ii) *rāhui*, which is concerned with the prohibitions and limitations of use of a particular place; iii) *tapu*, which invokes "respectful conduct" toward "spiritual qualities"; iv) *tapu me noa*, which denotes "sanctity .... [and] respectful behaviour" (tapu), and *me noa*, which means that once tapu is lifted from a place, it regains its usual state; and v) *tohu*, which "connotes the metaphysical or symbolic depiction of things" (New Zealand Legislation. 2021).

The other important legislation awarding the status of a legal person to natural ecosystems pertains to the Whanganui River. This Act, which is known as the "Te Awa Tupua (Whanganui River Claims Settlement) Act 2017", recognizes Te Awa Tupua (Whanganui River) as "an indivisible and living whole, comprising the Whanganui River from the mountains to the sea, incorporating all its physical and metaphysical elements" (*Section 12* of *Subpart 2* of the Act). Several intrinsic values of Te Awa Tupua (Whanganui River) are recognized that mark its essence. These values are: i) that "the River is the source of spiritual and physical sustenance"; ii) The tribes and subtribes (iwi and hapū, respectively) have inherent connections with the river, reflecting a Māori proverb that means "I am the River and the River is me"; and iii) "Te Awa Tupua is a singular entity comprised of many elements and communities". All these components work in unison to ensure "the health and well-being of Te Awa Tupua". These include the numerous small and large streams that flow into one another to form the river (*Section 13 of Subpart 2*). *Section 14* rules that "Te Awa Tupua is a legal person and has all the rights, powers, duties, and liabilities of a legal person". These rights, powers, and duties are to be carried out on its behalf by Te Pou Tupua, which is called the "human face" of the Whanganui River. The office of Te Pou Tupua comprises two persons: one nominated by the iwi (tribe), and another by the crown (New Zealand Legislation. 2022).

One implication of this judgment is that the river now owns itself, and ceases to be the property of anyone. It also indicates a widening of support – both legal and popular – for the growing demand for "Rights of Nature", or the "Rights of Mother Earth", which demands ecosystems to have legal rights "to exist, flourish and regenerate their natural capacities" (Biggs

2017). Moreover, since entities like Te Urewera or the Whanganui River have "all the rights, powers, duties, and liabilities of a legal person", they automatically possess intrinsic value, thereby lending an ecocentric touch to these and similar legislations.

The two New Zealand legislations are path-setting, with many positive aspects. Less than a year before the bill introducing the rights of the Whanganui River was formally passed in the New Zealand parliament in March 2017, Kathleen Calderwood posted an article on September 6, 2016, which mirrors the public perception of this unique legislation. This legislation as well as the 2014 Te Urewera Act raised hopes in the minds of many – especially the Māori – that these laws will enable natural entities like the Te Urewera and the Whanganui River or Te Awa Tupua to have their own authority. They could now think of their own well-being and be represented in court, albeit by authorized representatives of the Māori, who attach great cultural and spiritual significance to Te Urewera and Te Awa Tupua. They would also find the government on their side in raising issues that concern the health and well-being of these special natural entities. The Te Awa Tupua (Whanganui River) Act also showed the commitment of the government to reconcile with the Māori. This Act, therefore, rectifies some of the breaches of the Treaty of Waitangi that was signed between the Crown and the Māori chiefs in 1840 (Calderwood 2016). The background for the Whanganui River Act can be found in the signing of the Whanganui River Deed of Settlement (*Ruruku Whakatupua*) on August 5, 2014 between the Government of New Zealand and the Whanganui iwi. This marked a long struggle since the Treaty of Waitangi in 1840, because during the 1880s to 1920s, the Crown violated many provisions of the Treaty. These violations comprised, for example, the introduction of a steamer service, and extraction of minerals from the river-bed. These activities affected the fisheries and the general ecological conditions of the river, and degraded its cultural and spiritual values. These acts were done without consulting the Whanganui iwi, which has been pleading for redressal through its petitions to the Parliament and through several courts and the Waitangi Tribunal. The award of 80 million USD is provided by the bill to compensate these "actions and omissions" (Whanganui District Council 2023). Calderwood (2016) added that this judgment also honored the Māori perspective in which nature is personified, and the Earth is viewed as the Earth mother *Papatuanuku*. It also endorses the fact that neither the Māori nor the government or the Crown owns these lands, which cannot be owned by anyone.

9.4.1.1 Challenges Faced by Te Awa Tupua (Whanganui River)

Despite this general feeling of hope and optimism that the granting of personhood to Whanganui River will vindicate its mana (status) and maui (life force), the real challenges are not easy to overcome. These challenges are posed by pollution, contamination, and other anthropogenic disturbances from farmlands, forestry practices, damming, and urban-industrial development in the catchment of the river. While the headwaters of Whanganui and its tributaries are relatively free from human interferences, the river faces a number of harmful impacts in its downstream stretches. Even in the upstream areas, an energy company diverts a large proportion of the flow of the river and its tributaries for hydropower generation. This often results in downstream areas having reduced flow affecting their ecology. The Māori contend that though water from the river can be used for power generation, enough should be kept for sustaining the biodiversity and the cultural and spiritual values of the river. However, it will be very difficult to address this imbalance, because the company has the rights for diverting water at this rate till 2039, and the new Act cannot rectify this, because it does not have the power to nullify pre-existing laws and agreements. In the downstream areas, entry of sediments and bacterial load from the farmlands and human settlements pose problems for the health of the river. The plastic waste entering the river is collected by the canoe paddlers as they travel up the river, indicating rising awareness of the need to keep the river pollution-free. Another positive effect of the law is that the river now has guardians, who can speak for it in court, and though this protection may not be foolproof, it gives the river a better chance. Thus, though these laws cannot change the existing scenario overnight, they definitely raise hopes for a cleaner and healthier future for Te Awa Tupua (Lurgio 2019).

9.4.2 Granting of Rights of Nature and Personhood to Natural Systems in Latin American Countries

The framing of these two legislations in New Zealand, which uphold the principles of Earth jurisprudence, are not isolated initiatives confined to a few countries. For example, the highest court of Colombia granted rights of "protection, conservation, maintenance and restoration" to the river Rio Atrato, and its watersheds and tributaries, in November 2016. It is indeed significant to note the court's words that humans and the environment have a "profound relationship of unity and interdependence", which also reflects the "new socio-political reality" that seeks a "respectful transformation with the natural world and its environment". The court further observed that such a transformation has already taken place "with civil and political rights … economic, social and cultural rights … and environmental rights …" (Margil 2017). Another Latin American country that enacted a Rights of Nature law is Panama. The National Assembly of Panama passed Law 287, which "recognizes the rights of Nature and the duties of the State in relation to said rights" on February 24, 2022, jointly signed by Crispiano Adames Navarro, the President of the National Assembly of Panama, and Quibian T. Panay G., its Secretary General. *Article 1* of this law has the objective of recognizing "Nature as a subject of rights". It also aims to ensure the natural and legal "obligations ... [of] the State and all people .... to guarantee the respect and protection of these rights". *Article 3* stipulates that "The State must respect Nature in its existence integrally, due to its intrinsic value ...". Another eco-centric principle – "The Superior Interest of Nature" – is laid down in *Clause 1, Article 8*, which calls for "the special protection of fundamental rights of Nature, rooted in its intrinsic value". This is necessary because Nature is vulnerable to "human activities that may alter its ecological and vital cycles". In Clause 6, the Law states that "The cosmovision and the ancestral knowledge of the indigenous people of the country must be an integral part of the interpretation and application of the rights of Nature". Chapter II of this law elaborates on the "Rights of Nature". *Article 10* proclaims that "The State recognizes ... [several] minimum rights of Nature, that extend to all living beings, elements and ecosystems of which it is composed". These rights include:

1. "The right to exist, persist and regenerate their vital cycles".
2. "The right to diversity of life" and "ecosystems".
3. "The right to preservation of" the water cycle in a functional state, and the optimum "quantity and quality" of water.
4. "The right to the preservation of" air quality.
5. "The right to ... effective restoration of the life systems", which are "affected directly or indirectly by human activities.
6. "The right to exist free from contamination" by "toxic and radioactive residues generated by human activities. Nature has the right to live, exist and persist under its own framework of balanced development where ... the biological diversity or its components, may fulfil their function within it".

Article 12 lays down another stipulation, which is important from an ecocentric point of view. It says that "Nature has the right to preserve its biodiversity. Its living beings must be protected by law, independently from their utilitarian value to human beings". The law also binds the State and its institutions to several obligations toward Nature, which are "derived from the rights recognized in this Law" (*Article 16*) (National Assembly of Panama 2022).

9.4.3 Recognition of Rights of Nature in the USA

On September 17, 2016, the Ho-Chunk Nation in Wisconsin, USA, amended their tribal constitution to include the rights of nature. It became the first tribal nation of the United States to grant independent rights to nature. This amendment declares that "Ecosystems and natural communities within the Ho-Chunk Territory possess an inherent, fundamental, and inalienable right to exist and thrive" (CELDF 2023). The recognition of rights of nature (RoN) by the *Ho-Chunk* Nation is part of a tradition that has remained with them for long. The amendment, which was initially passed in 2016, was later revised and passed again in 2018. This constitutional amendment is still awaiting approval by the United States Bureau of Indian Affairs (Kauffman and Martin 2023).

The Ho-Chunk is not alone in advocating rights of nature. The *Ponca* Nation of Oklahoma have also introduced rights of nature laws. For the *Ponca*, it is "reviving sacred natural sites" which are threatened by an oil refinery, oil wells, and hydraulic fracturing and injection well sites. The *Ponca* resolution states that nature "gives sustenance and the opportunity for all people for all intellectual, moral, social and spiritual growth". It is the duty and responsibility of the people to not cause any harm to nature, because natural entities "are related to us and are a part of us". These principles have their origin in the ancient *Ponca* belief that "human beings are a part of Nature; that water is the container of all life, and that all life is the container of water". Under the rights of nature resolution, the *Ponca* Tribal Court is empowered to impose penalties for any crime against nature, which includes a prison term of up to one year and a fine of 5000 USD per day for each offense (Kauffman and Martin 2023).

The *Chippewa* Nation of Minnesota, who are also known as the *Ojibwa* or *Ojibwe*, and who call themselves *Anishinaabe*, are threatened by oil pipelines, and nickel and copper mines. Wild rice or *manoomin* is a culturally significant plant, food, and source of medicine to the *Ojibwe* tribe. In response to the withdrawal of protection to *manoomin*, the Minnesota *Chippewa* tribe formed its own "Tribal Wild Rice Task Force", which passed laws that recognize the rights of *manoomin* for the present and future generations. The resolution establishing these laws describes *manoomin* as "a gift to the *Anishinaabe* people from the creator or Great Spirit and an important staple of their diet for generations". It recognizes its right to "exist, flourish, regenerate and evolve", and its "inherent rights to restoration, recovery and preservation". It states that *manoomin* not only plays a very important part in "*Anishinaabe* culture, heritage and history", it is also an "integral part of the wetland ecosystems". The resolution also empowers the *Chippewa* to impose maximum fines for doing harm to *manoomin* under the Tribal Law (Kauffman and Martin 2023).

The *Yurok* Tribe, which is the largest tribe in California, has been living for centuries in the catchment of the Klamath River, California. They also have several sacred places along the Klamath. The *Yurok* passed a "Resolution Establishing the Rights of the Klamath River" in May 2019. The Resolution recognizes rights for the entire land of the *Yurok* Territory, including "the river, the trees, the salmon, elk, deer" and all plants and animals, and affirms that "all native species within and dependent on the

Klamath River ecosystem are vital to the cultural, legal, subsistence and economic interests" of *Yurok* people (Kauffman and Martin 2023).

Besides these tribal nations, some counties and local councils of the United States have also recognized rights of nature in different forms. In its "Coos County Bill of Rights", 2017, the Coos Commons Protection Council, Oregon, USA, has advocated "Rights of Natural Communities and Ecosystems to Thrive". It says that "ecosystems within Coos County, including but not limited to, forests, rivers, streams, wetlands, aquifers, near shore habitats, and intertidal zones possess the right to exist, flourish, and naturally evolve" without getting affected because of any unsustainable energy systems being constructed, sited or made operative in this County (Coos Commons Protection Council. 2019).

In response to a federal permit issued to the Pennsylvania General Energy (PGE) in 2013 to inject fracking waste into an abandoned well in Grant Township, Pennsylvania, USA, community members of the township adopted an ordinance in 2014 that recognized rights of nature and decided to block the injection wells. However, this ordinance was overturned by a federal judge in 2015 because it preempted state and federal law. To circumvent this obstacle, the Grant Township community adopted a Home Rule Charter in 2015, which is "a local municipal constitution that overrides the second-class status of a municipality in a US state" (Eco Jurisprudence Monitor 2023). The fundamental concept of home rule is that "the basic authority to act in municipal affairs is transferred from state law, as set forth by the General Assembly, to a local charter, adopted and amended by the voters". However, this "is a limited independence", which allows "shifting of responsibility for local government from the State Legislature to the local community", and the municipalities are still "subject to restrictions found in the United States and Pennsylvania constitutions and in state laws applicable to home rule municipalities" (Pennsylvania Department of Community and Economic Development 2020). The Home Rule Charter adopted by the Grant Township declares in its Section 104 that

> All residents of Grant Township, along with natural communities and ecosystems within the Township, possess the right to clean air, water, and soil, which shall include the right to be free from activities which may pose potential risks to clean air, water, and soil within the Township, including the depositing of waste from oil and gas extraction.

Section 105 of the Charter declares that "All residents of Grant Township possess the right to the scenic, historic, and aesthetic values of the Township, including unspoiled vistas and a rural quality of life". This includes "the right .... to be free from activities which threaten scenic, historic, and aesthetic values, including from the depositing of waste from oil and gas extraction".

Rights of nature are enunciated in Section 106, which states that "Natural communities and ecosystems within Grant Township, including but not limited to, rivers, streams, and aquifers, possess the right to exist, flourish, and naturally evolve". The Pennsylvania State Department of Environmental Protection went to court in 2017, arguing that the Charter was in violation of state law, under the provisions of which injection wells, was granted permission. The trial court invalidated the Charter, though this ruling was reversed by an appellate court. This court ruled that Pennsylvania state's constitution guaranteed "peoples' right to clean air, pure water, and the preservation of the environment", and the act of permitting injection wells went contrary to this guarantee. At this juncture, the PGE appealed to the court to request "judgment against the town without trial". Decision on this was pending till 2022.

9.4.4 Recognition of Rights of Nature in the UK

In the UK, the Green Party of England and Wales became the first national party to adopt rights of nature in its policies. This resolution was taken in its 2016 spring conference at Harrogate in Yorkshire, UK. The conference resolved that:

> Rights of Nature is a new way of conceptualizing our relationship with nature. What we are looking at here is no less than a fundamental paradigm shift away from the toxic perception of nature as an object to be consumed.

This resolution of the Green Party, UK and Wales, was preceded by the adoption of a similar decision by the Scottish Greens or the Scottish Green Party, in 2015. According to a Party representative, granting rights to nature "is an idea whose time has come" (The Ecologist 2016).

9.4.5 Granting of Personhood to Ganga and Yamuna Rivers in India

In a judgment delivered on March 20, 2017, the High Court of Uttarakhand state in India observed that "Rivers Ganges and Yamuna are worshipped by Hindus … [and are] sacred and revered. The Hindus have a deep spiritual connection with Rivers Ganges and Yamuna". The judgment invoked the concept of "Juristic Person" or a "juridical person", which "connotes recognition of an entity to be in law a person which otherwise it is not". This means that it is not a "natural person" but "an artificially created person". The judgment also referred to the definition that "A legal person is any subject-matter other than a human being to which the law attributes personality". The judgment refers to a well-known book of jurisprudence (Paton 1967), which says that entities other than an individual human being, such as a group of human beings, or a fund, or an idol may be accorded the status of a legal personality. Thus, a group of persons forming a corporation, or an institution, such as a church, a hospital, a university, a library, a charitable fund, or a trust estate may be regarded as a legal person. Therefore, "for the purpose of jurisprudence", a person may be defined as "any entity (not necessarily a human being) to which rights and duties may be attributed". The Court also observed that "for a bigger thrust of socio-political-scientific development" it is often necessary to designate "a fictional personality to be a juristic person". "A juristic person, like any other natural person ..... is also conferred with rights and obligations ... [but] acts ... only through a designated person". "As in the case of minor a guardian is appointed", so is this guardian or designated person legally authorized to represent the juristic person. The rivers Ganga and Yamuna "have provided both physical and spiritual sustenance ... from time immemorial ... [and] are breathing, living and sustaining the communities from mountains to sea".

> Accordingly, ... the Rivers Ganga and Yamuna, all their tributaries, streams, every natural water flowing with flow continuously or intermittently of these rivers, are declared as juristic / legal persons / living entities having the status of a legal person with all corresponding rights, duties and liabilities of a living person in order to preserve and conserve river Ganga and Yamuna. The Director NAMAMI Gange [flagship program of Government of India for clean Ganga], the Chief Secretary of the State of Uttarakhand and the Advocate General of the State of Uttarakhand are hereby declared *persons in loco parentis* [in the place of a parent] as the human face to protect, conserve and preserve Rivers Ganga and Yamuna and their tributaries. These officers are bound to uphold the status of Rivers Ganges and Yamuna and also to promote the health and well-being of these rivers.
>
> *(words within box brackets are author's words)*

The judgment also "banned forthwith" "mining in river bed of Ganga and its highest flood plain area" (Live Law News Network 2017). However, this judgment was stayed by the Supreme Court (the highest court in India) in delivering judgment on a petition from the Uttarakhand Government that the high court verdict raised several legal questions and administrative issues. For example, since the rivers flow through several states, only the Center could frame rules for their management. The ruling also raised questions like whether the victim of a flood in the rivers can sue the state for damages and also about whether the state and its officers will be liable in case of pollution in the rivers in another state through which it flows (Express News Service 2017).

9.5 The Beginnings of a New Paradigm?

The *harmony with nature* initiative of the United Nations might be said to have been initiated on April 22, 2009, with a speech by Evo Morales Ayma, the president of Bolivia, and the proclamation of April 22 as the "International Mother Earth Day" by the UN General Assembly. This was followed by the first UN General Assembly resolution on *Harmony with Nature* on December 21, 2009, followed by 12 more annual resolutions (second to thirteenth) between 2010 and 2022. These resolutions asked the member states to convene interactive dialogues to discuss different issues relating to human-nature relationships; to explore a more ethical and non-anthropocentric basis of human–Earth relationships; to recognize the "fundamental interconnections between humanity and nature"; to promote "harmony with the earth as found in indigenous cultures"; to appeal to "citizens and societies" to reexamine the ways in which they interact with the natural world; to discuss the "relationship between harmony with nature and the protection of biological diversity"; and to evolve a "non-anthropocentric or Earth-centered paradigm" (United Nations 2023).

9.6 Conclusions

Reducing the anthropocentric bias of environmental policies, legislations and governance systems by accommodating biocentric and ecocentric considerations may help in achieving lasting solutions to the various environmental problems. Earth jurisprudence has taken up the cause of nature and the environment for the sake of the whole suite of biodiversity, and the biosphere in its entirety. However, its true purpose will be defeated if rights given to nature remain simply as window dressings or more alarmingly, a sugar coating under the cover of which unrestrained destruction of nature can be continued. The world community has to serve its role as a watchdog to ensure the true manifestation of Earth jurisprudence.

10

ECOCENTRISM BEYOND EARTH

Ethics of Space Exploration

10.1 Introduction

Human colonization of space is no longer confined to the domains of science fiction and fantasy. Teams of astronauts have lived in space stations on a rotation basis to understand the potential effects of long-term habitation in space. Conditions in other planets such as Mars have been simulated in certain remote and extremely inhospitable areas of the Earth. Experiences of scientists and officials living in such areas have also yielded valuable information on the possible effects of living in space. The challenges facing the future space colonists include exposure to harmful radiations, living in low-gravity environments for long periods of time, and the psychological stress of loneliness, isolation in a limited space, and detachment from family and friends. The possible effects of isolation during prolonged stay in hostile environments, such as that found on Mars, are being studied by NASA scientists in Devon Island, Canada. Devon Island is a large uninhabited island, which is located in the northernmost extremity of Canada in the Arctic Circle. It has an extremely hostile climate with winter temperatures dropping sometimes to –50° Celsius. This, and the island's barren terrain and remoteness, makes it suitable for serving as an "analog Martian environment", which provides ideal conditions for simulating a prolonged stay in space stations or on a planet like Mars (Perez 2019; Keeter 2021). These, and other preparations, are being made to make human forays into space, which are expected to be more frequent in the coming years, to be safe and comfortable. However, human exploration, habitation, and resource exploitation in space also raise some ethical questions.

DOI: 10.4324/9781003481058-10

10.2 Sustainability in Space Exploration

Besides the establishment of "lunar colonies", there is an increasing interest in lunar exploration for using the Moon as a resource base for meeting growing human demands. Newman (2015) has raised the question about the prudence of such programs undertaken without giving due consideration to the issue of sustainability. The entry of private enterprises in lunar missions, while laying emphasis on exploitation of resources and profitability of investments, may neglect to ensure adequate environmental safeguards and compromise sustainability. It is, therefore, necessary to develop appropriate international agreements to ensure sustainability in lunar mining and other forms of activities on the surface of the Moon. The Outer Space Treaty (OST) of 1967, which has been discussed in some detail in Section 10.7.1, laid down the principle that space exploration and use shall be for the benefit of the entire humankind, with freedom for scientific investigations. However, such explorations should avoid "harmful contamination" of the Moon and other bodies and the Earth. All these prospects and problems of the exploration of and resource extraction from the Moon and other celestial bodies bring into foreground the issue of geodiversity protection and conservation not only on the surface of the Earth but also on that of the Moon, Mars, and any other planets, satellites, or asteroids. IUCN defines geodiversity as "the variety of rocks, minerals, fossils, landforms, sediments and soils, together with the natural processes that form and alter them. It includes past and present geological and geomorphological features". A geoheritage is "those elements and features of the Earth's geodiversity … that are considered to have significant value for intrinsic, scientific, educational, cultural, spiritual, aesthetic, ecological or ecosystem reasons and therefore deserve conservation". It is "a legacy from the past" that needs to be conserved at the present and left in that state for the future generations (Crofts et al. 2020, 6). The Digne-Les-Bains Declaration – Declaration of the Rights of the Memory of the Earth – states that mankind and all life are bound in intricate and indispensable linkages with the Earth, which sustains the diversity of life, and serves as the connection between humans and all the other forms of life (ProGeo 1991). This concept of geodiversity protection should now be extended to include the celestial bodies. The Moon especially is likely to be exposed to a host of anthropogenic activities in the near future. The moon regolith can become a rich source of rare metals and helium-3, the mining of which would involve extensive disturbances on the surface of the moon. Thus, increased human presence on the Moon poses environmental and ethical challenges, besides the obvious technological ones. It has, therefore, been proposed that several areas on the Moon should be protected as "international multiuse land reserves" (Capper 2022). In view of the exploitative and oppressive connotations of the word "colonization" in human history, Bill Nye of the science education television program *Bill Nye the Science Guy* has proposed to replace this term with that of "settlement". It is also suggested by several scholars that replacing this term could also enable humans to think differently and orient their attitudes and actions towards a responsible, safe, and sustainable exploration of space and utilization of extraterrestrial resources (Wall 2019a).

10.3 Relevance of Geodiversity and Geoethics in Space Exploration and Resource Utilization

Geoethics lays down principles in the context of framing appropriate moral guidelines for human interactions with the Earth system. It calls for geoscientists and the general public to be more aware of the emergence of humans as powerful geological agents, which in turn makes it their ethical responsibility to ensure sustainable use of natural resources, reduce rampant modification of natural landforms, and prevent pollution of ecosystems (Di Capua and Peppoloni 2019). These principles of geoethics are also applicable in the context of the exploitation of resources of the Moon, Mars, and other celestial bodies. This chapter also argues for the incorporation of more ecocentric ideals in the tenets of geoethics to enable it to emerge from its largely anthropocentric confines. This is especially important in view of the International Union for Conservation of Nature (IUCN) advocating for protection and conservation of geodiversity "for ethical reasons", or in other words, for its "intrinsic value", and not merely because of its usefulness for human society. The objective of geoconservation is to conserve geodiversity "just because it is there: *for its own sake*". This motto of geoethics is in conformity with ecocentric principles. Therefore, an ecocentric geoethics could provide the normative guidelines for natural resource exploration and exploitation on the Earth and even beyond it.

10.4 Human Settlements in Space

NASA plans to land the first woman and first person of color on the Moon through its Artemis missions. It is also having a highly ambitious plan to initiate the establishment of a lunar base around 2028. The European Space Agency (ESA) is also working to establish a permanent lunar settlement – the Moon Village, though it is expected to take more time. The Moon Village Association, which is an NGO created in 2017, with its headquarters in Vienna, Austria, has 33 institutional members from 65 countries. Its objective is to serve as a forum for promoting cooperation and interaction between a large number of stakeholders interested in realizing the concept of a settlement on the lunar south pole near the rim of the Shackleton Crater. The Moon Village of ESA or the lunar base of NASA is also envisaged as a research hub, destination for travelers, and training base for future launches to Mars and beyond (Parks 2019).

Besides the establishment of lunar or Martian settlements, space tourism is also coming of age with three space travel companies – Space Adventures, Blue Origin, and Virgin Galactic – in operation as of 2021. With the possible spurt in space tourism in the coming years, several legal and ethical issues also emerge, which need to be addressed (Li 2023).

10.5 Mining and Other Human Interventions on the Moon, Mars, and the Asteroids

The concept of In-Situ Resource Utilization (ISRU) has gained popularity because it can reduce the cost of establishment and maintenance of human settlements on the Moon, Mars, and other celestial bodies by utilizing the mineral resources found in the Moon or Mars. Another objective of mining the lunar or Martian surface is to extract the resources and utilize them for various purposes on the Earth. The Moon, Mars, several other planets and their moons, and the asteroids contain a large number of mineral resources. For example, the Australian Space Agency and NASA have a project that aims to send a rover to the Moon to collect lunar rocks that could be a potential source of oxygen for future lunar colonies or settlements. Though the atmosphere of the Moon does not contain oxygen, a large amount of oxygen is trapped in lunar regolith, which contains approximately 45% oxygen. It is estimated that the top ten meters of lunar regolith could provide oxygen to the Earth's entire population for around 100,000 years. However, producing oxygen from regolith is an energy-intensive process, and would also require transportation of industrial equipment to the Moon (Grant 2021). In a recently published article, *The Guardian* reported that NASA aims to start mining activities in the Moon to initiate soil excavation by 2032. NASA's Artemis missions will be used to develop commercial prospects and attract private players. The prospective customers include rocket companies, which can use Moon's resources as fuel (Oladipo *and agency* 2023).

Studies on the geology of the Moon show the presence of water in the rocks, and in the polar ice. Water can have multiple uses: for drinking and growing food crops, as well as for splitting into hydrogen and oxygen, and the former can be used as a fuel. Helium-3, which is abundant in the lunar regolith, finds its use in future energy technologies such as nuclear fusion. Rare Earth metals are essential in the manufacture of smartphones, computers, several medical equipment, and also in emerging technologies. Regolith can also be explored for constructing building blocks for Moon habitations. Understanding its properties would also enhance the ability to use "lava tubes", that is the caves and hollows for developing these as safe and secure human habitats. However, mining in Moon poses stiff challenges in terms of the necessity of inventing innovative mining technologies, and overcoming the problem of transportation of equipment. Off-Earth mining equipment to be used on the Moon or Mars would, therefore, have to be smaller in size, but more powerful. The question of the legality of lunar mining by a country or a group of countries, damage to lunar environment, and ethical issues also stand in the way of mining the lunar surface (Hall 2020; Martin 2022).

The Martian resources that could be extracted and utilized include the following: i) Martian clay for making bricks; ii) quartz and iron-free opaline silicate rocks to manufacture glass and fiberglass; iii) perchlorates as antifreeze and as a potential energy source; iv) water ice to yield water and hydrogen; v) magnesium; vi) geolites; vii) nickel; viii) and others. Therefore, mining of Martian soil and rocks for building and sustaining human habitations holds lucrative prospects, and USGS has already launched a project to explore the feasibilities and potential of mineral resources on Mars for future colonization (USGS 2023).

Asteroid mining for mineral resources is currently beyond feasibility because of the astronomically high cost and other technical issues. However, if these issues could be resolved, asteroids could provide a mind-boggling amount of various minerals, including gold. It is estimated that mining ten asteroids which are closest to the Earth and are resource-rich could yield a 1.5 trillion US dollar profit. It is estimated that the asteroid 16 Psyche contains "US $ 700 quintillion worth of gold", which if divided among every person on the Earth would be "about US $ 93 billion" (Yarlagadda 2022). However, question remains whether influx of such large quantities of gold would affect its value.

Another approach towards large-scale human intervention is that of "terraforming" (transform a planet to resemble Earth to support human life). Various possible mechanisms have been suggested for terraforming Mars. The US billionaire and SpaceX CEO Elon Musk has suggested detonating a large number of very small low fallout nuclear fusion explosions to create artificial suns, which would vaporize the polar ice of Mars to generate greenhouse gases like water vapor and carbon dioxide into the Martian atmosphere. This is expected to warm up the planet from its present freezing conditions and create Earth-like conditions, without making it radioactive (Wall 2019b; Letenyei 2023). However, many scientists are skeptical of this idea and say that nuclear explosions might also generate large amounts of dust and particulates to produce a "nuclear winter". Another alternative put forward by NASA scientist Jim Green is to place a giant magnetic shield in orbit between Mars and the Sun in order to prevent solar wind and radiation from stripping the Martian atmosphere. Once the stripping is stopped, the planet is expected to restore its atmosphere and transform itself to a warmer place (Bennett 2017). Casey J. Handmer of the Terraform Industries would like to launch "small scale solar sails" into Low Earth Orbit (LEO) where they will fly themselves to Mars, reflect sunlight onto its dark side, and warm it up over a period of ten years or so at a nominal cost of $ 10 billion. Handmer is conscious of the ethical issues associated with assuming control to this extent, though he overlooks it considering the human propensity to modify the environment (Handmer 2022).

All these ideas, plans, and programs view outer space and its celestial bodies as storehouses of valuable resource, and meant exclusively for the welfare of humans. Further, they tend to exert human control over their landforms to modify and deface them, transform their atmosphere, and make them subservient to human needs. They are envisaged to serve as "back-up" repositories of essential resources after humans have exhausted their reserves in the Earth through reckless exploitation, destroying in the process the ecosystems and geographical/geological formations that harbor them and their associated biodiversity, and stress the environment with pollution and toxic contamination.

10.6 Ethical and Legal Issues in Space Mining and Space Settlements

It is evident from the preceding discussion that outer space has become the new frontier for colonization, exploitation, and commercialization, processes that have already been initiated, and which are expected to intensify in the coming years. The projected human activities in outer space, some of which have been reviewed here, naturally raise ethical questions. Laura Poppick posed several relevant questions in view of rapid advancements in the ability of humans in space travel and possible colonization of Mars and other celestial bodies. One question she raised is whether humans should take offensive measures if they are threatened by Martian life forms such as microbes. Poppick also pointed out that the existing treaties and guidelines are mainly meant to ensure "the safety of humans and the scientific evidence of life" on Mars or other celestial bodies, but they do not deal with the protection of the environment of these bodies. Ethical questions are likely to arise when abiotic resources are exploited in Mars or other planets/satellites. This is a complex issue that is not easy to resolve, because new challenges may crop up as new exoplanets and such other bodies are explored. This calls for a flexible yet efficient and dynamic approach (Poppick 2017).

Amalyah Hart raised the question of the rights of mining and extraction of the Moon's mineral and other resources. These lunar resources are attracting different countries as well as private players to explore the possibility of lunar mining. Besides the Moon, Mars and the asteroids are also in the scheme of things for mining their resources in future. With this scenario developing fast, Hart raised the important legal-ethical issue of the legal rights of the countries for lunar resource extraction. The Outer Space Treaty (OST) of 1967, which provides an "overarching legal framework" for human activities in space, declares that no country is entitled to any claim of sovereignty of the outer space, including the Moon and other celestial bodies (see Section 10.7.1 for more details). Therefore, no country can claim ownership of any sorts over any area on the Moon. However, the Treaty allows

> exploration and use of outer space, including the Moon and other celestial bodies ... for the benefit and in the interests of all countries .... Without discrimination of any kind, ... [with] free access to all areas of celestial bodies.
>
> *(words within box brackets are author's words)*

However, many experts feel that the Treaty is somewhat ambiguous in this regard, because if certain resources are extracted by a country, that automatically excludes access to those resources for everybody else. Furthermore, I would also like to add here that such access – though assuring of "equality" in principle – would actually remain confined to a handful of technologically developed countries capable of mobilizing manpower, scientific know-how, and technology, for exploiting the resources in outer space and its celestial bodies. Since this Treaty (or any other agreement) does not prescribe any benefit-sharing mechanism, benefits from the resources in the Moon, other celestial bodies, or anywhere in outer space would be garnered by a few countries.

Besides OST, there are other treaties such as the 1979 UN "Agreement Governing the Activities of States on the Moon and Other Celestial Bodies", and the 2020 Artemis Accords tabled by the Department of state, USA, and the National Aeronautics and Space Administration (NASA) (see Section 10.7.2 and 10.7.3 for more details). However, the Moon Agreement was not signed by the USA, Russia, and China, while as of 2023, only 27 nations have signed the Artemis Accords, which do not include Russia and China. The 2021 draft Declaration of the Rights of the Moon adopts a different position by recognizing the rights of the Moon as "an autonomous natural system". These facts suggest that a consensus on the extraction of resources from the Moon, the other celestial bodies, and outer space in general, has so far not been reached (Hart 2023). More details on these treaties, accords, and declarations are given in Section 10.7.

The nature of space ethics that we need to follow is also a debated issue. An anthropocentric approach would readily allow humans to enjoy the benefits of space colonization. However, it can be observed that "some of our Earth-ethics ... deep ecology for example" have turned their attention to the nonhuman environment as well. This shift has prompted the inclusion of "off-Earth biospheres" and raised the question of recognizing intrinsic value in "extraterrestrial microbial life and abiotic environments". A "non-anthropocentric ... perspective" would view the "commodification of space resources .... as instrumentalizing them", and robbing them of their intrinsic value (Segobaetso 2018). I would like to point out here that Segobaetso speaks of the classical and "strong" anthropocentric ethic, which presently is followed by very few people – including policy makers and administrators – worldwide. A weaker version that assigns a stewardship role to humans, and incorporates more prudence and sustainability in its norms, is currently more acceptable to the world community. Nevertheless, even the weaker version still retains a strong flavor of anthropocentrism, putting human interests ahead in the name of achieving "sustainable" development, especially citing the need for less developed nations to achieve industrial and economic growth. Therefore, when we talk about ethical implications in planetary protection, a question that automatically arises is whether anthropocentric ethical approaches can provide adequate planetary protection, and whether they are morally justifiable, or whether the ethical considerations need to delve into biocentric or even ecocentric philosophies. The latter approach is broader and all-encompassing, and therefore, able to include in its fold celestial bodies that are totally devoid of life. On the other hand, Segobaetso refers to Wilfred Beckerman and Joanna Pasek (Beckerman and Pasek 2010, as cited in Segobaetso 2018), who argued that anthropocentrism is the most appropriate ethic with regard to outer space and celestial bodies, because when humans protect nature, they place themselves outside nature, and regulate their own activities to save nature from human depredation. On the other hand, if humans consider themselves as a part of nature, then the changes wrought on nature – whether by humans or by natural processes – are simply a part of the evolutionary process. Recognizing intrinsic value in nature – both in the Earth and in outer space – would raise questions on the method of defining the "interests" of outer space and trading off these "interests" against that of humans. Further, humans would also have to decide whether the moral rights of humans are greater than or equal to that of outer space. Segobaetso also cited David Goldsmith and Robert Ries, who pointed out that adoption of non-anthropocentric ethics such as Deep Ecology will be opposed by most utilitarians who would like to see more commercial activities in space that serve the interests of societal development. Under a non-anthropocentric model, commercial activities and explorations would have to work within limits set by stringent regulations for protecting extraterrestrial life forms (in places where they exist) as well as the non-living components of the environment. Such measures would consequently stymie human progress (Goldsmith and Ries 2008, as cited in Segobaetso 2018).

These opposing arguments raise the question of whether utilitarian interests induce us to ignore or even relax our planetary protection measures. The coming decades are likely to witness an intensive search for life on Mars, as well as explorations into the oceans of the icy moons, for example, Europa of Jupiter and Enceladus of Saturn, which are located at the outer zone of the solar system beyond what is known as the frost line. Scientists think that the oceans beneath the icy surface of these satellites may contain life. While this is an exciting prospect, such explorations run the risk of possible contamination of these pristine environments with Earth microbes. This would necessitate giving a very high priority to ethical concerns, which may lead to review and revision of the existing principles of planetary protection. Such measures are also in keeping with Article IX of the Outer Space Treaty (OST) of 1967 (see more on OST in Section 10.7.1), which stipulates that explorations conducted on the Moon and other celestial bodies shall take care to prevent harmful contamination of their environments. Taking these precautions is extremely important because if these environments are contaminated with Earth microbes, then even in the event of life being detected in them, it would be very difficult to ascertain whether those life-forms are native to that particular celestial body, or they have been introduced from the Earth. That there are possible threats of earthly contamination is vindicated by the fact that 298 strains of extreme bacteria (extremophiles are bacteria that can survive hostile conditions such as intense radiation, vacuum pressure, temperature extremes, and microgravity) could withstand the sterilization procedure followed in the "clean rooms" of the European Space Agency. Therefore, it is quite possible that some terrestrial microbes may already exist in a dormant state in Mars, though it is unlikely that they have reproduced there (Gronstal 2018). The threat of contamination can be further ascertained from the recovery of a number of fungal taxa such as *Aspergillus*, *Mucor*, *Candida*, and others, and bacterial taxa such as *Micrococcus*, *Staphylococcus*, *Streptococcus*, etc. from both crewed and robotic spacecrafts. Stringent planetary protection measures are, therefore, justified in order to effectively safeguard the pristine nature of the environment of the celestial bodies, and guard against possible contamination from Earth microbes and any other biological contaminants accidentally spilled during space missions. However, besides adopting and implementing scientific measures to prevent contamination, this issue needs to be addressed by adopting more holistic models that extend beyond science to include "ethical implications and responsibility [of humans] to avoid harmful impact on potential indigenous biospheres". This also calls for planetary protection protocols to be elevated from mere prevention of contamination to protection of the entire environment of planets and other celestial bodies along with the adoption of ethical norms and principles (Nicholson et al. 2009).

Daniel Munro strongly advocated for "an acceleration and expansion of space ethics" commensurate with the increase in space explorations and "private commercial activities". It is essential to ensure that while framing human activities in space, strategic and commercial interests should not override ethical considerations. Munro also pondered over the question of how to prioritize "scientific, strategic and commercial interests" in the event of conflicts arising among them. He suggested that while resolving issues concerning commercial activities such as mining that pose the threat of contamination, scientific considerations should hold precedence over strategic and commercial interests. Mining should be totally prohibited if the situation so demands. Most importantly, space ethics "should be insulated from ideology, interests and bias". Success of space ethics at the policy level is also handicapped by the fact that while academic ethicists are free to raise the relevant questions, their influence at the policy implementation level is minimal. On the other hand, operational practitioners have the necessary influence and resources, but have to work within institutional constraints. The role of elected policy makers has mostly remained far from satisfactory, and has been driven by strategic or commercial interests (Munro 2022).

John D. Rummel mentions some ethical issues that crop up during space exploration:

1.   Moral obligations towards living systems outside Earth;
2.   Possible relevance of ecocentrism in outer space; and
3.   Which nonhuman entities in space have intrinsic value?

The measures that should be adopted by the Committee on Space Research (COSPAR) during space exploration include:

1.   Environmental impact assessment for human activities in space;
2.   "An intergovernmental mechanism" for enforcing regulations for space exploration and exploitation; and
3.   "Designation, establishment and monitoring" of "Planetary Parks" (Rummel 2017).

Seven such planetary parks have been proposed in Mars, which are areas where landing of unmanned crafts will not be allowed. These parks will have regulations to ensure the maximum level of protection (Cockell and Horneck 2004). Similarly, during lunar resource exploitation, certain areas should be protected as "international multiuse land reserves". These may include the Malapert Massif near the Moon's south pole, several craters near the north pole and on the far side, and the landform perceived as a woman (old woman on a spinning wheel in parts of India and Bangladesh), "man, toad or rabbit in the moon" (Capper 2022).

Mars has a fragile environment, and though it seems that it does not harbor intelligent life, presence of microbial life cannot be ruled out yet. Besides, Mars may have minerals of great value to humans, and already there are plans and proposals for mining minerals in Mars (see Section 10.4 for more details). Based on our experiences of the devastating environmental impacts of mining on Earth, the impact of such activities on the Martian landscape is not likely to be very different (Dirks 2021). In this context, the principles of geoethics could be of great value while planning mining operations on Mars, the Moon, and any other celestial body including asteroids (see Section 10.8 for more discussion on this).

10.7 Existing Treaties and Agreements on Space Exploration, Settlements, and Resource Extraction

10.7.1 The Outer Space Treaty

The "Treaty on Principles Governing the Activities of States in the Exploration and Use of Outer Space, including the Moon and Other Celestial Bodies", or the "Outer Space Treaty" (OST), remains the baseline document on rules governing outer space exploration and use. Adopted on December 19, 1966 by the UN General Assembly, this Treaty was signed by the Russian Federation, the United Kingdom, and the United States of America, and came into force in October 1967. *Article I* of this treaty stipulates that all countries, regardless of the state of their "economic or scientific development", shall be entitled to the benefits of "The exploration and use of outer space, including the Moon and other celestial bodies". Outer space is "the province of all mankind", which all countries are free to explore and use on an equitable basis. *Article II* declares that no country shall have any claim of sovereignty on outer space, including the Moon and other celestial bodies, "by means of use or occupation, or by any other means". "States parties to the treaty" shall conduct studies on exploration and use by "maintaining international peace and security and promoting international cooperation and understanding" (*Article III*). No country should place any "nuclear weapons or any other kinds of weapons of mass destruction" in the Earth's orbit, and refrain from establishing "military bases, installations and fortifications", and "the testing of any type of weapons", and conducting "military maneuvers" (*Article IV*). *Article IX* stipulates that exploration of outer space including the Moon and other Celestial Bodies shall be conducted "so as to avoid their harmful contamination and also adverse changes in the environment of the Earth resulting from the introduction of extraterrestrial matter". Furthermore, all activities in the outer space shall be reported to the "Secretary-General of the United Nations" (*Article XI*) (UNOOSA 2008).

10.7.2 Agreement on the Moon

Besides the OST, the UN General Assembly also adopted the "Agreement Governing the Activities of States on the Moon and other Celestial Bodies" in its 89th Plenary Meeting in 1979. This Agreement recognizes the importance of the Moon "in the exploration of outer space", aims to prevent it "from becoming an area of international conflict", and is aware of "the benefits which may be derived from the exploitation of the natural resources of the Moon and other celestial bodies". The Agreement – which also applies to other celestial bodies besides the Moon – emphasizes the use of their resources exclusively for peaceful purposes in an atmosphere of international cooperation (*Article 1 and 2*). Establishing military base or fortifications and putting in lunar orbit nuclear weapons or other devices of mass destruction are prohibited (*Article 3*). "The exploration and use of the Moon shall be the province of all mankind" (*Article 4*); and "There shall be freedom of scientific investigation on the moon by all States Parties without discrimination of any kind" (*Article 6*). *Article 7* of this agreement states that

> In exploring and using the Moon, States Parties shall take measures to prevent the disruption of the existing balance of its environment, by its harmful contamination through the introduction of extra-environmental matter or otherwise. States Parties shall also take measures to avoid harmfully affecting the environment of the Earth through the introduction of extraterrestrial matter or otherwise.

The States Parties shall also have to "notify … [the UN Secretary-General] … in advance of all placements by them of radioactive materials on the Moon and of the purposes of such placements". Thus, this Article, on the one hand, introduces restrictions on contaminating the Moon surface, though on the other hand, it allows any State to put radioactive materials, albeit with the permission of the UN Secretary-General. All countries are also allowed to conduct their exploration "anywhere on or below its surface", subject to fulfilling certain conditions (*Article 8*). The States Parties may set up "manned and unmanned stations on the moon" (*Article 9*); and "The moon and its natural resources are the common heritage of mankind, … and is not subject to national appropriation" (*Article 11*) (UNOOSA 2008).

10.7.3 The Artemis Accords

One of the objectives of the Artemis Accords, 2020, on the exploration of space and space resources exploitation is to "usher in a new era of exploration" that will lead to gaining of "global benefits of space exploration and commerce". Some of the defining principles of these Accords include ensuring peaceful purposes, transparency through "broad dissemination of information", interoperability in terms of "common exploration infrastructure and standards", "registration of space objects", sharing of scientific data, mutual rendering of "emergency assistance", preservation of "outer space heritage", "deconfliction of space activities", and safe disposal of orbital debris. One very important principle pertains to the utilization of space resources to benefit humankind. The "Accords" were initially signed by the representatives of Australia, Canada, Italy, Japan, Luxembourg, United Arab Emirates,

United Kingdom, and the United States of America. Nineteen more states have so far signed the Accords, taking the total number of signatories to 27 (US Department of State and NASA 2020).

10.7.4 Ethical Issues

All these treaties, agreements, and accords lay great emphasis on the exploration of the vast resources of the Moon and other celestial bodies for global benefits, though they also emphasize the need for ensuring that such activities do not cause any adverse impact on their environment. Thus, they are essentially anthropocentric, albeit with some environmental safeguards. Though basically remaining reductionist and anthropocentric, the International Science Council (ISC) has entrusted the Committee on Space Research (COSPAR) with providing international standards for planetary protection. This Committee is of increasing relevance because of the launching of more and more ambitious and innovative projects by a larger number of nations and private and/or commercial agencies using advanced technologies including robotics involving "sample return". These developments make it imperative to ensure the protection of "the pristine environments" that are subject to human activities, as well as the safety of the Earth and its biosphere. The COSPAR recommendations are "not legally binding", though these are "internationally endorsed". The COSPAR Planetary Protection Policy (PPP) aims to ensure that any space mission sent by a country or any private agency does not contaminate the planet or satellite with biological material, especially microorganisms transported from the Earth ("forward contamination"). It also takes care that extraterrestrial material brought back from a planet or a satellite ("backward contamination") does not contaminate the biosphere. While making an assessment of the risks of biological contamination of the Earth during missions to Venus, Mars, and other small bodies, COSPAR found that Venus, including its clouds, does not pose any discernible risks in terms of planetary protection. However, in the case of Mars, "the presence of a biological hazard in Martian material cannot be ruled out". In the case of small bodies, there is no apparent risk of forward contamination, though this should be finalized on a case-to-case basis (Coustenis et al. 2023).

In contrast to the anthropocentric approach in the treaties and agreements described above, the "Declaration of the Rights of the Moon" puts forward distinctly ecocentric concepts for the preservation of the ecological and existential integrity of the Moon. This declaration by "the people of Earth" acknowledges "the unique, intact, interconnected lunar environments and landscapes" and "the ancient, primordial relationship between Earth and the Moon"; and the "deep cultural and spiritual meaning" of the moon for humans. Being aware of the human impact on the Earth causing "ecosystem collapse", biodiversity loss, and "global climate change", it intends to avoid the repetition of similar adverse impacts on the "natural systems and ecosystems of the Moon". This Declaration, therefore, calls for giving "The Moon – which consists of but is not limited to: its surface and subsurface landscapes including mountains and craters, rocks and boulders, regolith, dust, mantle, core, minerals, gases, water, ice, boundary exosphere, surrounding lunar orbits, cislunar space" – the status of "a sovereign natural entity in its own right", without any claim of ownership by any individual, nation, or any other entity of Earth. The Moon, therefore, has the right to retain its natural cycles and "ecological integrity", and remain unpolluted. It also has "the right to be defined as a self-sustaining, intelligent, cohesive, intact lunar ecosystem", and should be kept in a peaceful state of existence, without being affected "by human conflict and warfare" (AELA 2021).

10.8 Geoethics and Its Relevance in Planetary Protection

As discussed in Section 10.3, geoethics intend to frame appropriate moral guidelines for human interactions with the Earth system, and ask geoscientists as well as the general public to be more aware of the emergence of humans as powerful geological agents. The environmental issues raised and discussed in the preceding sections imply that its principles can be extended to the exploration and exploitation of space resources for prescribing the ethical guidelines towards ensuring sustainable use of natural resources, reducing rampant modification of natural landforms, and preventing pollution of ecosystems both in the Earth as well as the Moon, Mars, and other celestial bodies. Bobrowsky et al. (2018) affirmed that "Geoethics aims to provide a framework of values" in order to enable geoscientists to work in an ethically responsible manner. As defined by the International Association for Promoting Geoethics (IAPG),

> Geoethics consists of research and reflection on the values that underpin appropriate behaviors and practices, wherever human activities interact with the Earth system … [it] deals with the ethical, social and cultural implications of geoscience knowledge, education, research, practice and communication, providing a point of intersection for Geosciences, Sociology, Philosophy and Economy... [and] is a tool to influence the awareness of society regarding problems related to geo-resources and geo-environment.
>
> *(Di Capua and Peppoloni 2019; words within*
> *box brackets are author's words)*

Geoethics can make possible the adoption of a multidisciplinary scientific and technological approach in the exploration of and research on the Moon and Mars, accompanied by due attention to issues in ethics and scientific integrity. Geoethics, which is concerned with both "scientific and societal" aspects, now finds a broader application, with its principles being extended to the subject fields of "planetary geology and astrobiology". Integration of geoethical principles into these "planetary" disciplines would enable them to incorporate the concept of geodiversity and geoheritage (see Introduction section) as proposed by the IUCN (Crofts et al. 2020). The adoption of geoethical principles also necessitates the framing of appropriate regulations on disturbances caused during "robotic and manned planetary missions", and establishment of "wilderness / planetary parks". Data from Earth locations which are called "terrestrial analogs" should also be utilized while framing regulations (Martinez-Frias et al. 2010). "A terrestrial analog is a field site, [on Earth] which bear an analogy or similarity in some way to other planetary bodies of the solar system". For example, the Atacama Desert in Argentina and Chile, and Lake Vostok – a subglacial lake in Antarctica – are regarded as terrestrial analogs of Mars, and Jupiter's moon Europa, respectively. The factors taken into consideration for determining analogy include "physicochemical conditions, such as dryness, low water activity, mineralogy, or chemical composition of the environment" (Gǿmez 2011). The Lanzarote and Chinijo Islands Geopark is another such analog site. This Geopark comprises an area of over 2500 sq km in the larger Lanzarote Island and a group of small islands called Chinijo Islands, which are a part of the Canary Islands in the Atlantic Ocean, off the west coast of Africa. The geopark area, especially the island of Lanzarote, is used as analog sites for conducting studies that would be useful in the exploration of Mars and the Moon. Various geological, geochemical, mineralogical, and astrobiological studies of volcanic rocks, and the interactions of these rocks with water and their mineralization processes, are conducted in these islands. Other studies include the testing of lunar and Martian rovers and other equipment, and the fabrication of regolith simulants (Martinez-Frias et al. 2017). These analog sites, therefore, suggest that it would be relevant to apply geoethical principles in planetary protection.

10.9 Principles of Geoethics

Geoethical guidelines have been developed by incorporating some ethical principles already laid down in general ethics and environmental ethics. However, these principles, when adopted in geoethics, deal with aspects like "deficiency of mineral resources, their exhaustibility, non-renewability", and the fact that they not only belong to the present, but also the future generations. Some of these principles may be summarized as follows:

1. The Earth should be viewed as having its own value, "and not as an object of industrial impact";

2. "Principle of sympathy", which treats abiotic nature including the "geo-logical environment" from the perspective of "its interests";
3. Principle of interrelation, which states that "no geosystems, planetary or local" exist in isolation, and are interrelated with other systems;
4. "Principles of harmony and balance of interests", while exploiting "min-eral resources" and "intruding into the geological environment";
5. "Principle of geodiversity conservation";
6. "Principle of responsibility", which implies that development should take into account the needs and aspirations of future generations;
7. "Precautionary principle", for taking preventive measures, even when the risks are not fully proven scientifically; and
8. "Principle of integration", which calls for introduction of these ethi-cal norms into "laws, standards and rules of conduct of nations of the world" (Nikitina 2016). Peppoloni and Di Capua (2016) list some ethical values of geosciences such as "respect for natural systems and dynamics when designing interventions on the environment; protection and enhancement of geodiversity"; and "promotion of sustainability", to conserve sufficient energy sources and natural resources for future generations. The protection of geodiversity is also linked with cul-tural values, because geodiversity and geoheritage also possess cultural value, especially for many indigenous communities all over the world. Therefore, geoethics enable geoscientists to act for the betterment of society, and to make it "more respectful towards the environment".

These principles are also reflected in the Cape Town Statement on Geoethics (CTSG) of 2016, which was drafted during the 35th International Geological Congress at Cape Town, South Africa, organized by the International Association for Promoting Geoethics (IAPG). The objective of CTSG was to draw the attention of geoscientists to the importance of adopting ethical principles in the study and practice of geology. The CTSG aims to inculcate some "fundamental values of geoethics" such as "honesty, integrity, trans-parency, reliability, … competence", knowledge sharing and communica-tion, and "spirit of cooperation and reciprocity". Besides these basic values, CTSG promotes "respecting natural processes and phenomena, … when planning and implementing interventions in the environment"; protection and enhancement of geodiversity and geoheritage; ensuring economic and social sustainability; and promoting geo-education and outreach to develop and improve "geohazard prevention and mitigation, and environmental pro-tection" models to increase "societal resilience and well-being". Under the "geoethical promise", CTSG requires geoscientists to "be fully respectful of Earth processes", while working in the field of geoscience (Di Capua et al. 2017). The International Association for Promoting Geoethics (IAPG) has promoted publications that contain "reflections on the intersection between geosciences and humanities, social sciences and economics". Thus, geoethics is serving as an academic arena where scholars from both within and outside the geosciences can contribute towards the enrichment of geoethical thoughts and concepts. It is necessary for geosciences to understand the nature of the relationship between "human reality" and nonhuman nature, thereby acting as "a cultural bridge between science, society and nature". Some geoscientific practices in the past had contributed to the creation and intensification of ecological unsustainability and ecosystem degradation. Geoethics is to renew a "sense of wonder at the knowledge of nature and its beauty" (Di Capua et al. 2021).

10.10 Planetary Protection and Conservation: Arguing for an Ecocentric Approach

At present, space exploration is largely guided by anthropocentric views of seeing the human species as an entity distinct from the nonhuman world, and is on the threshold of "conquering the Universe". Andrea Owe wants an ecocentric approach to replace the anthropocentric view, though Owe's version of "ecocentric space expansion" suggests either terraforming or "constructing Earth-like environments from scratch" as "promising methods" for making ecocentric expansion a reality. Furthermore, Owe would like to "bring terrestrial nature into space nature", and, if prioritization between the two is called for, to give priority to terrestrial nature (Owe 2023). Needless to say, these approaches go against the cardinal principle of ecocentrism of respecting natural creations, whether biotic or abiotic. Such attitudes promote the idea of subjugation that so characterized the Western colonization of large chunks of the Eastern world, with the accompanying damages inflicted on the ecosphere. The "terrestrial nature" of Owe could be considered as somewhat equivalent to Western systems and values which were imposed on the colonized parts of the world. The proposed establishment of earthly ecospheres throughout spacetime with humans in the role of their "morally responsible stewards" reflects a philosophical viewpoint of "weak anthropocentrism", as opposed to ecocentrism, which views humans as "partners" or participants", or even reaching the state of "union with nature" (Zweers 1994).

Charles S. Cockell and Gerda Horneck suggested that humans must develop an ethic, and a policy backed by that ethic, on the exploration and resource utilization in Mars. Such an ethic is essential because it is possible that life exists on Mars, and there are scientific as well as ethical reasons for leaving the Martian life forms undisturbed. However, even if Mars does not harbor life, it would be ethical to control human activities on Mars due to its value in itself, irrespective of the presence of life there. Cockell and Horneck (2006), therefore, advocated for establishing protected planetary parks in certain areas on Mars that have "particular scientific, aesthetic, and historical value". Similar Parks could be established on the Moon, and other celestial bodies wherever necessary. They also advanced different arguments justifying the establishment of these Parks, which are similar to those behind the establishment of wilderness areas on the Earth. Some of these arguments have an anthropocentric and utilitarian basis, such as setting aside wilderness areas as a mark of the "civilized" status of the human species, which is capable of thinking about the impact of its own activities. The other utilitarian reasons include preserving such areas for the benefit of the future generations, the possibility of receiving "unknown and indirect" benefits from Martian land in future. Besides these arguments, it is also suggested that Mars, or the Moon, or any other celestial body has intrinsic value, that is a value for its own sake (Cockell and Horneck 2006). Thus, this value is regardless of any present or future use value of these celestial bodies for humans. This argument has an ecocentric basis and shows how ecocentrism can have its influence even beyond the Earth. Leonard David had quoted Gerda Horneck – a leading researcher in space ethics and policy – who is of the opinion that a planetary park system would still allow commercial activities in areas not protected through these parks, and accommodate "both utilitarian and intrinsic-value arguments ... and would allow us to express a respect for other worlds" (David 2013).

10.11 Conclusions

The discussion in the preceding sections of this chapter has revealed the plans and programs undertaken by national agencies as well as private corporations for establishing human settlements on the Moon and Mars, and explorations for extracting mineral and other resources on these, other planets and satellites, and the asteroids. All such activities need to be guided by some international regulations and standards. The Outer Space Treaty (OST) of 1967 – though vague and deficient in certain respects – can be said to comprise the first step by denying ownership rights over any natural entity in space or parts of it by any nation, company, or individual. However, it does not prohibit or even regulate resource exploration and extraction from space, provided that these activities do not cause any harmful contamination or any other adverse impact on the Moon and other bodies, and the Earth. This Treaty was signed by 110 States, and signed but yet to be ratified by another 89 nations. Besides, there was an agreement on the Moon and other celestial bodies in 1979, which has not been signed by the USA, Russia, and China, which are the major space-faring nations. Furthermore, this Agreement has been signed only by four nations, with another 18 being party to it. This Agreement also largely has an anthropocentric approach like the OST. On the other hand, the "Declaration of the Rights of the

Moon" has a distinct ecocentric approach, but is yet to be recognized or accepted by any state or international agency. In this prevalent scenario, the principles of geoethics, which have the scope of being extended as planetary ethics, provide several ecocentric guidelines for protecting the Moon, Mars, and other celestial bodies that are likely to bear the onslaught of human activities in the future. Thus, ecocentric ideals for outer space explorations have found voice – though feeble at present – in some declarations and in the principles of geoethics.

EPILOGUE

This book tries to show that recognition of intrinsic value in nature and beliefs in biocentric/ecocentric worldviews have remained distinct traits of the human mind, perhaps since the early dawn of human civilization. Different shades of ecocentric thoughts and beliefs, which are either explicit or implicit in different forms of animisms, totemism, and tree and nature worships, have persisted in humans in various forms: in nature poetry and other creative writings, in Deep Ecological and ecocentric thoughts, certain streams of conservationist approach, various religious and spiritual philosophies, altruism toward animals and nature, and a general sense of love, respect, and awe for nature. Despite compassion and reverence for nature persisting among many humans, worldviews professing denial of the rights of nature and not giving it any moral consideration have held sway in the sphere of public policies, which appear to suggest that ecocentrism (in all its variants) has failed to leave its mark in the minds of people in power and capable of decision-making, at almost all levels. More often than not, however, the same person who has a "soft" corner for nature and is ready to leave nature some space, even at the risk of reduced personal gains, moves to the other side after assuming a position of responsibility. A school teacher who loves the trees on a plot of spare land in her/his school, and is averse to cutting them down, might agree to do so, when s/he serves as the principal, and would have to nod to the irrefutable logic of the expansion and "development" of the institution. Finally, almost everyone has to bow down to the powerful God of development. It is ironical that on the one hand, believing in the existence of spiritual force or intrinsic value in nature, or worshiping nature is tantamount to primitive and irrational fetishism, though on the other hand, worshiping the God of development is regarded as representing rational and scientific thinking. Ecocentrism could never become the dominant worldview, though it refused to die or be totally exterminated.

All ecocentrists are keen to see a paradigm shift from anthropocentrism to ecocentrism occur in all human relationships with nature. And though it is ideal for this transformation to take place voluntarily in people's minds through a spontaneous ecocentric recognition of nature in one's conscience, in reality, various vested interests prevent people from making this "jump". It is, therefore, necessary that the adherents of ecocentric ideals try to have these principles adopted in environmental policies at international, national, and local levels, and become an integral part of our legal framework and jurisprudence.

The reverence for nature, and an inclination to view humans as a part of nature and a part of the natural processes, might have emanated from the realization that human survival depended on resources provided by nature; resources which were otherwise not available or could become difficult to obtain during specific periods. Hence, there was a need to propitiate nature, either directly through nature worship or through the guardian deities for a particular resource or group of resources. In other words, recognition of intrinsic value sometimes had extrinsic roots, which were forgotten with time. Irrespective of its basis, intrinsic value in nature ensured its protection and prevented over-exploitation and abuse to a great extent. There was a holistic and a hermeneutic understanding of nature – it was placed on a pedestal equal in stature to or higher than the status of humans.

The facts and concepts presented in this book also try to suggest that ecocentrism is more than an academic construct or a philosophical postulate. It has its roots running into the great depths of human tradition with global manifestations, and we may not be wrong to speculate that it is a trait integrated into our genetic core.

The word *dharma* in Sanskrit or *dhamma* in Pali (ancient Indian languages) is often understood and used as a synonym for religion. The answer to the question "what is your *dharma*?" would typically be "it's Hinduism" (or Islam or Sikhism or Christianity, as the case may be). However, *dharma* or *dhamma* has a much wider connotation than religion, and has in fact several shades or facets of meaning. It denotes the eternal reality, and the fundamental properties of animate or inanimate entities. Thus, a particular metal or an alloy would have its specific *dharma*, just as a particular human being or an animal would have their own *dharma* as well. In its deepest and most profound sense, *dharma* is a diffuse yet distinctly delineated concept.

Dharma binds us all – humans, plants, animals and non-living materials, ecosystems, and landscapes. In that sense, it has an ecological dimension. More than four decades ago, Barry Commoner in his book *The Closing Circle* put forward the first law of ecology: "everything is connected with everything else". Just as we might perceive ourselves to be connected with each other through energy flow or material cycles, in a cultural context we could also visualize *dharma* connecting everything in the Earth.

If *dharma* is understood as conformity to duty, a question to ask is toward whom do we have these duties? Should the long list also include that toward the plants and animals and to the ecosystems where they (and we) live? If nature is understood as a property integral to our minds, then does love and compassion for nature represent a property that we all possess to varying degrees? Exceptions do exist, but aberrations are common in nature. They are there to remind us of or reiterate the truth. Only when there is the realization of this *dharma* of interconnectedness, altruism can well up and inundate our senses, our very existence – altruism that like the concentric ripples on still water can transcend our kin, our caste, race, religion, even species, to embrace the others and the whole biosphere.

Viewed in this light, E.O. Wilsons' biophilia or an extended version of it, which can be termed as ecophilia, is a *dharma* that all humans possess, either consciously or sub-consciously. In many cases, this innate property may be masked by cultural influences to be either expressed as biophobia, or as an affinity for man-made, tailored "second nature". Yet, a total bio or ecophobia is hard to find, since most people like at least some part of nature – a dog-hater might like kittens, or flowers in bloom, or rivers or mountains.

Even if *dharma* is understood as religion, did our early ancestors not worship nature or its components – the elements like thunder and rain, or rivers and mountains, or plants such as *Ficus* and others, or the myriads of animals? This was the primal *dharma*, the religion that was born from instinct, from the intuitive desire to survive and thrive and go ahead. The ecologist Paul Shepard had said: "the beauty and complexity of nature are continuous with ourselves" and "the world is a being, a part of our own body" (Shepard and McKinley 1969 cited in Cramer and Van Den Daele 1985). This book tries to suggest that ecocentrism is a basic human property, which – given the right setting – can facilitate a paradigm shift in the dominant worldview. The growing need for such a transformation may also enable us to successfully overcome or at least manage the environmental crisis that we are experiencing today.

REFERENCES

Adamatzky, Andrew. 2022. "Language of fungi derived from their electrical spiking activity." *Royal Society Open Science* 9: 211926. https://doi.org/10.1098/rsos .211926

Alves, Bruna. 2023. "Global plastic waste flow 2019." *Statista 2023.* https://www .statista.com/statistics/1357641/plastic-waste-lifecycle-worldwide/ [Accessed 03 July 2023].

Altman, Sam, Greg Brockman, and Ilya Sutskever. 2023. *Governance of Superintelligence.* OpenAI. https://openai.com/blog/governance-of -superintelligence [Accessed 29 June 2023].

Anderson, Mark. 2012. "New Ecological Paradigm (NEP) scale." *The Berkshire Encyclopedia of Sustainability: Measurements, Indicators, and Research Methods for Sustainability.* www.berkshirepublishing.com

Animal Ethics. 2021. *Sentience.* https://www.animal-ethics.org/sentience-section/ [Accessed 13 July 2022].

Aratani, Y., T. Uemura, T. Hagihara, K. Matsui, and M. Toyota. 2023. "Green leaf volatile sensory calcium transduction in *Arabidopsis*." *Nature Communications* 14: 6236: 1–16. https://doi.org/10.1038/s41467-023-41589-9.

Asiaticus. 1912. "The rise and fall of the indigo industry in India." *The Economic Journal* 22 (86): 237–247.

Attfield, Robin. 1981. "The good of trees." *Journal of Value Inquiry* 15: 35–54.

Attfield, Robin. 1999. *The Ethics of the Global Environment.* Edinburgh: Edinburgh University Press.

Attfield, Robin. 2009. "Biocentrism." In *Encyclopedia of Environmental Ethics and Philosophy*, Volume 1: Abbey to Israel, edited by J. Baird Callicott and Robert Frodeman (Editors in Chief), 97–100. Detroit: Macmillan Reference USA, A Part of Gale, Cengage Learning.

Australian Earth Laws Alliance (AELA). 2021. *Declaration of the Rights of the Moon.* Australian Earth Laws Alliance. https://www.earthlaws.org.au/moon -declaration/ [Accessed 19 July 2023].

Australian Earth Laws Alliance (AELA). 2023. *Rights of Nature.* Australian Earth Laws Alliance. https://www.earthlaws.org.au/aelc/rights-of-nature/ [Accessed 06 July 2023].

Baard, Patrik. 2019. "The goodness of means: Instrumental and relational values, causation, and environmental policies." *Journal of Agricultural and Environmental Ethics* 32: 183–199. https://doi.org/10.1007/s10806-019-09762-7

Baluška, František. 2016. "Should fish feel pain? A plant perspective." *Animal Sentience* 3 (16): 2016.023. https://doi.org/10.51291/2377-7478.1052

Baluška, František, and Stefano Mancuso. 2009. "Plant neurobiology: From sensory biology, via plant communication, to social plant behavior." *Cognitive Processing* 10 (Suppl. 1): S3–S7.

Baluška, František, and Stefano Mancuso. 2013. "Root apex transition zone as oscillatory zone." *Frontiers in Plant Science* 4 (Article 354): 1–15. https://doi.org/10.3389/fpls.2013.00354

Baluška, František, Simcha Lev-Yadun, and Stefano Mancuso. 2010. "Swarm intelligence in plant roots." *Trends in Ecology and Evolution* 25 (12): 682–683.

Baluška, František, Stefano Mancuso, Dieter Volkmann, and Peter Barlow. 2009. "The 'root-brain' hypothesis of Charles and Francis Darwin." *Plant Signaling & Behavior* 4 (12): 1121–1127. https://doi.org/10.4161/psb.4.12.10574

Barnard, Alan. 1981. "Universal kin categorization in four Bushman societies." *L'Uomo* V (2): 219–237.

Basu, Buddhadeb. 1957. *Prabandha Sankalan.* Calcutta: Dey's Publishing.

Batavia, Chelsea, and Michael Paul Nelson. 2017. "For goodness sake! What is intrinsic value and why should we care?" *Biological Conservation* 209: 366–376.

Baxter, Brian. 2005. *A Theory of Ecological Justice.* London and New York: Routledge.

Bayram, Selma Aydin. 2012. "On the role of intrinsic value in terms of environmental education." *Procedia – Social and Behavioral Sciences* 47: 1087–1091.

Bellarsi, Franca. 2009. "The challenges of nature and ecology." *Comparative American Studies An International Journal* 7 (2): 71–84.

Bennett, Jay. 2017. "NASA considers magnetic shield to help Mars grow its atmosphere." *Popular Mechanics*, March 2. https://www.popularmechanics.com/space/moon-mars/a25493/magnetic-shield-mars-atmosphere/ [Accessed 23 July 2023].

Bergson, Henri. 1999. "Creative evolution." In *The Biosphere and Noosphere Reader*, edited by Paul R. Samson and David Pitt, 57–60. Taylor and Francis e-Library Edition, 2002. London and New York: Routledge.

Berry, Thomas. 1995. "The viable human." In *Deep Ecology for the 21st Century: Readings on the Philosophy and Practice of the New Environmentalism*, edited by George Sessions, 8–18. Boston: Shambhala.

Berry, Thomas. 1999. *The Great Work: Our Way into the Future.* New York: Bell Tower.

Biggs, Shannon. 2017. "Rivers, rights and revolution: Learning from the Māori." In *Rights of Nature & Mother Earth: Rights-Based Law for Systemic Change*, edited by Shannon Briggs, Tom B.K. Goldtooth, and Osprey Orielle Lake, 24–26. Oakland: Movement Rights, Women's Earth & Climate Action Network, Indigenous Environmental Network.

Biggs, Shannon, Osprey Orielle Lake, and Tom B.K. Goldtooth. 2017. "The challenge is upon us: Climate chaos or communities of cooperation." In *Rights of Nature & Mother Earth: Rights-Based Law for Systemic Change*, edited by Shannon Briggs, Tom B.K. Goldtooth, and Osprey Orielle Lake, 3. Oakland: Movement Rights, Women's Earth & Climate Action Network, Indigenous Environmental Network.

Billings, Linda. 2020. "Earth, life, space: The social construction of the biosphere and the expansion of the concept into outer space." In *Social and Conceptual Issues in Astrobiology*, edited by Kelly C. Smith, and Carlos Mariscal, 239–C14.

N80. New York: Online Edition, Oxford Academic, June 18. https://doi.org/10
.1093/oso/9780190915650.003.0014

Birch, Jonathan. 2017. "Animal sentience and the precautionary principle." *Animal Sentience* 16 (1): 2017.017. https://doi.org/10.51291/2377-7478.1200

Birch, Jonathan. 2022. "The search for invertebrate consciousness." *NOÛS* 56: 133–153. https://doi.org/10.1111/nous.12351

Bird-David, Nurit. 1990. "The giving environment: Another perspective on the economic system of gatherer-hunters." *Current Anthropology* 31 (2): 189–196.

Bird-David, N. 1999. "'Animism' revisited: Personhood, environment, and relational epistemology." *Current Anthropology* 40 (S1): S67–S91. Special Issue Culture – A Second Chance?

Bloomfield, M., trans. 1897. *Hymns of the Atharva-Veda: Together with Extracts from the Ritual Books and the Commentaries.* Sacred Books of the East, Volume 42. https://ia600207.us.archive.org/34/items/hymnsofatharvave00bloo/hymnsof atharvave00bloo.pdf [Accessed 30 September 2021].

Bobrowsky, Peter, Vincent S. Cronon, Giuseppe Di Capua, Susan W. Kieffer, and Silvia Peppoloni. 2018. "The emerging field of geoethics." In *Scientific Integrity and Ethics with Applications to the Geosciences*, edited by L.C. Gunderson, 1–42. Hoboken, NJ and Washington, D.C.: John Wiley and Sons, Inc. and American Geophysical Union.

Bogert, Jeanne Marit, Jacintha Ellers, Stephan Lewandowsky, Meena M. Balgopal, and Jeffrey A. Hurvey. 2022. "Reviewing the relationship between neoliberal societies and nature: Implications of the industrialized dominant social paradigm for a sustainable future." *Ecology and Society* 27 (2): 7. https://doi.org/10.5751 /ES-13134-270207

Bolle, Kees W., Jonathan Z. Smith, and Richard G.A. Buxton. 2020. "Myth." *Encyclopedia Britannica*, November 3. https://www.britannica.com/topic/myth [Accessed 2 August 2022].

Bookchin, Murray, and Dave Foreman. 1991. *Defending the Earth: A Debate.* Montréal/New York: Black Rose Books. https://theanarchistlibrary.org/ library/murray-bookchin-and-dave-foreman-defending-the-earth-a-debate.pdf [Accessed 01 August 2022].

Bookchin, Murray. 1987a. "Social ecology versus deep ecology: A challenge for the ecology movement." *Green Perspectives: Newsletter of the Green Program Project* nos. 4–5. Anarchy Archives: An Online Research Center on the History and Theory of Anarchism. http://dwardmac.pitzer.edu/Anarchist_Archives/ bookchin/socecovdeepeco.html

Bookchin, Murray. 1987b. "Thinking ecologically: A dialectical approach." *Our Generation* 18 (2): 3–40.

Bose, Buddhadeva. 1948. *An Acre of Green Grass: A Review of Modern Bengali Literature.* Calcutta: Papyrus.

Braithwaite, Victoria. 2010. *Do Fish Feel Pain?* Oxford: Oxford University Press.

Braithwaite, Victoria A. and L.O.E. Ebbeson. 2014. "Pain and stress responses in farmed fish." *Revue Scientifique et Technique – Office International des Épizooties* 33 (1): 245–253.

Braithwaite, Victoria, and Paula Droege. 2016. "Why human pain can't tell us whether fish feel pain." *Animal Sentience* 3 (3). https://doi.org/10.51291/2377 -7478.1041

Brassington, Iain. 2012. "The concept of autonomy and its role in Kantian ethics." *Cambridge Quarterly of Healthcare Ethics* 21: 166–176.

Brazil, Jon. 2000. "Dreamtime superstore: Encountering Australian aboriginal beliefs." *Third Text* 14 (50): 61–72. https://doi.org/10.1080/09528820008576837

Brøgger, Fredrick Chr. 2007. "Anthropocentric nature lover: Annie Dillard and the transcendentalist tradition in American nature writing." *American Studies in Scandinavia* 39 (2): 29–40.

Broom, Donald M. 2016. "Fish brains and behaviour indicate capacity for feeling pain." *Animal Sentience* 3 (4). https://doi.org/10.51291/2377-7478.1031

Brown, Culum. 2015. "Fish intelligence, sentience and ethics." *Animal Cognition* 18 (1): 1–17. https://doi.org/10.1007/s10071-014-0761-0

Brown, Culum. 2016. "Fish pain: An inconvenient truth." *Animal Sentience* 3 (32): 2016.058. https://doi.org/10.51291/2377-7478.1069

Bshary, Redouan. 2011. "Machiavellian intelligence in fishes." In *Fish Cognition and Behavior*, Second Edition, edited by Culum Brown, Kevin Laland, and Jens Krause, 277–297. Chichester, West Sussex: Wiley-Blackwell.

Buber, Martin. 1923. *I and Thou*, First South Asian Edition, 2005. London and New York: Continuum.

Buber, Martin. 1947. *Between Man and Man*. London and New York: Routledge 2002 edition.

Buddharakkhita, A. 2009. *Mettā: The Philosophy & Practice of Universal Love*. Bangalore: Reprint-March 2014. Buddha Vachana Trust, Mahabodhi Society.

Burbery, Timothy J. 2012. "Ecocriticism and Christian literary scholarship." *Christianity and Literature* 61 (92): 189–214.

Burdon, Peter. 2011. "The great jurisprudence." *Southern Cross University Law Review* 14: 1–18.

Byrne, Richard W., and Andrew Whiten. 1997. "Machiavellian intelligence." In *Machiavellian Intelligence II: Extensions and Evaluations*, edited by Andrew Whiten and Richard W. Byrne, 1–23. Cambridge: Cambridge University Press.

Calderwood, Kathleen. 2016. "Why New Zealand is granting a river the same rights as a citizen." *ABC Radio National*. https://www.abc.net.au/radionational/programs/sundayextra/new-zealand-granting-rivers-and-forests-same-rights-as-citizens/7816456 [Accessed 06 July 2023].

Callicott, Baird J. 1999. *Beyond the Land Ethic: More Essays in Environmental Philosophy*. New York: University of New York Press.

Callicott, Baird J. 2006. "Conservation values and ethics." In *Principles of Conservation Biology*, Third Edition, edited by Martha J. Groom, Gary K. Meffe, and C. Ronald Carroll, 111–135. Sunderland: Sinauer Associates, Inc.

Calvo, Paco. 2017. "What is it like to be a plant?" *Journal of Consciousness Studies* 24 (9–10): 205–227.

Calvo, Paco, and František Baluška. 2015. "Conditions for minimal intelligence across eukaryota: A cognitive science perspective." *Frontiers in Psychology* 6: Article 1329. https://doi.org/10.3389/fpsyg.2015.01329

Calvo, Paco, and Anthony Trewavas. 2020. "Physiology and the (neuro)biology of plant behaviour: A farewell to arms." *Trends in Plant Science* 25 (3): 214–216.

Calvo, Paco, František Baluška, and Anthony Trewavas. 2021. "Integrated information as a possible basis for plant consciousness." *Biochemical and Biophysical Research Communications* 564: 158–165. https://doi.org/10.1016/j.bbrc.2020.10.022

Calvo, Paco, Vaidurya Pratap Sahi, and Anthony Trewavas. 2017. "Are plants sentient?" *Plant, Cell & Environment* 40: 2858–2869. https://doi.org/10.1111/pce.13065

Calzadilla, Paola V., and Louis J. Kotzé. 2018. "Living in harmony with nature? A critical appraisal of the rights of Mother Earth in Bolivia." *Transnational Environmental Law* 7 (3): 397–424. https://doi.org/10.1017/S2047102518000201

Cancela da Fonseca, Jorge P. 2000. "On Vernadsky's biosphere." *Web Ecology* 1: 86–96.

Capper, Daniel. 2022. "What should we do with our Moon?: Ethics and policy for establishing international multiuse lunar land reserves." *Space Policy* 59: 101462. https://doi.org/10.1016/j.spacepol.2021.101462

Cartwright, Mark. 2017. "Kami." *World History Encyclopedia*, April 04. https://www.worldhistory.org/Kami/ [Accessed 30 September 2021].

CELDF. 2023. *Ho-Chunk Nation General Council Approves Rights of Nature Constitutional Amendment.* Press Release: Community Environmental Legal Defense Fund (CELDF) News, September 18. https://celdf.org/2016/09/press-release-ho-chunk-nation-general-council-approves-rights-nature-constitutional-amendment/ [Accessed 14 July 2023].

Chakrabarty, Dipesh. 2009. "The climate of history: Four theses." *Critical Inquiry* 35 (2): 197–222.

Chan, K.M.A., P. Balvanera, K. Benessaiah, M. Chapman, S. Diaz, E. Gómez-Baggethun, R. Gould, N. Hannahs, K. Jax, S. Klain, G.W. Luck, B. Martín-López, B. Muraca, B. Norton, K. Ott, U. Pascual, T. Satterfield, M. Tadaki, J. Taggart, and N. Turner. 2016. "Why protect nature? Rethinking values and the environment." *PNAS* 113 (6): 1462–1465.

Chapple, C.K. 2001. "The living cosmos of Jainism: A traditional science grounded in environmental ethics." *Daedalus* 130 (4): 207–224.

Chase, S. 1991. "Introduction: Whither the radical ecology movement?" In *Defending the Earth: A Debate*, 7–19. Montréal/New York: Black Rose Books. https://theanarchistlibrary.org/library/murray-bookchin-and-dave-foreman-defending-the-earth-a-debate.pdf [Accessed 01 August 2022].

Chaudhuri, Sukanta. 2011. "Tagore looks East." Working Paper # 2. Singapore: Nalanda Sriwijaya Centre. https://www.iseas.edu.sg/images/pdf/nscwps2.pdf

Chittka, Lars, and Fei Peng. 2013. "Caffeine boosts bees' memories." *Science* 339: 1157–1159.

Clark, John P. 2010. "A dialogue with Arne Naess on social ecology and deep ecology (1988–1997)." *The Trumpeter* 26 (2): 20–39.

Clark, John. 1996. "How wide is deep ecology?" *Inquiry* 39 (2): 189–201. https://doi.org/10.1080/00201749608602416

Cockell, Charles, and Gerda Horneck. 2004. "A planetary park system for mars." *Space Policy* 20 (4): 291–295. https://doi.org/10.1016/j.spacepol.2004.08.003

Cockell, Charles S., and Gerda Horneck. 2006. "Planetary parks – Formulating a wilderness policy for planetary bodies." *Space Policy* 22: 256–261.

Cohen, Marc S. 2004. "Aristotle on the soul." https://faculty.washington.edu/smcohen/320/psyche.htm. Last updated on 09/23/2016 [Accessed 9 September 2022].

Commoner, Barry. 1971. *The Closing Circle: Nature, Man and Technology.* New York: Alfred A. Knopf.

Convention on Biological Diversity. 2022. "Conference of the parties to the convention on biological diversity, fifteenth meeting – Part II." CBD/COP/15/L.25. https://www.cbd.int/doc/c/e6d3/cd1d/daf663719a03902a9b116c34/cop-15-l-25-en.pdf

Coos Commons Protection Council. 2019. *Coos County Bill of Rights.* https://cooscommons.org/ordinance/ [Accessed 14 July 2023].

Costanza, Robert, John Cumberland, Herman Daly, Robert Goodland, and Richard Norgaard. 1997. *An Introduction to Ecological Economics.* St. Lucie Press and ISEE.

Coustenis, Athena, Niklas Hedman, Peter T. Doran, Omar Al Shehhi, Eleonora Ammannito, Masaki Fujimoto, et al. 2023. "Planetary protection: An international concern and responsibility." *Frontiers in Astronomy and Space Science* 10: 1172546. https://doi.org/10.3389/fspas.2023.1172546

Cramer, J. and W. Van Den Daele. 1985. "Is Ecology an 'alternative' natural science?" *Synthese* 65: 347–375.

Crofts, R., J.E. Gordon, J. Brilha, M. Gray, J. Gunn, J. Larwood, V.L. Santucci, D. Tormey, and G.L. Worboys. 2020. *Guidelines for Geoconservation in Protected and Conserved Areas.* Best Practice Protected Area Guidelines Series No. 31, IUCN. Gland: Switzerland. https://portals.iucn.org/library/sites/library/files/documents/PAG-031-En.pdf [Accessed 05 September 2023].

Crutzen, Paul J. 2002. "The geology of mankind." *Nature* 415: 23.

Cusack, Carole. 1998. "Sacred groves and holy trees." https://openjournals.library .usyd.edu.au/index.php/SSR/article/download/12108/11253 [Accessed 22 November 2022].

Damasio, Antonio R. 1998. "Investigating the biology of consciousness." *Philosophical Transactions of the Royal Society of London B* 353: 1879–1882.

Damasio, Antonio R. 1999. *The Feeling of What Happens – Body and Emotion in the Making of Consciousness.* New York: Harcourt Brace and Company.

Darwin, Charles. 1871. *The Descent of Man, and Selection in Relation to Sex.* London: J. Murray, Photo reproduction published by Princeton University Press, 1981.

Darwin, Charles. 2009. *The Power of Movement in Plants.* Edited by Francis Darwin. Cambridge: Cambridge Library Collection, Cambridge University Press (Originally published by London: John Murray, 1880).

Das, Jibanananda. 1942. *Banalata Sen.* Calcutta: Kabita Bhavan.

Das, Jibanananda. 2010. *Ruposhi Bangla*, First Internet Edition, compiled and edited by Nabiul Afroz.

Das, K.C. 1978. "An approach to Assam vaishnavism in the light of the Upanishads." In *Sankaradeva: Studies in Culture*, edited by B.P. Chaliha, Second Edition, 1998. Assam: Srimanta Sankaradeva Sangha. https://www.atributetosankaradeva.org/ approach.pdf [Accessed 27 August 2023].

Datiles, Marianne Jennifer, and Pedro Acevedo-Rodriguez. 2014. "*Indigofera tinctoria* (true indigo)." *Invasive Species Compendium*, CABI International. https://www.cabi.org/isc/datasheet/28613#tosummaryOfInvasiveness [Accessed 14 July 2022].

David, Leonard. 2013. "Planetary parks could protect space wilderness." *Space.com*, January 17. https://www.space.com/19302-planetary-parks-space-wilderness -protection.html [Accessed 27 July 2023].

Dawkins, Marian S. 2000. "Animal minds and animal emotions." *American Zoologist* 40: 883–888.

Dawoud, Galal Eldin Hussein Eltahir. 2017. *Emily Dickinson's Poetry from an Ecofeminist Perspective.* Master's Thesis, Al-Neelain University, Sudan.

Deb, D., and K.C. Malhotra. 2001. "Conservation ethos in local traditions: The West Bengal heritage." *Society and Natural Resources* 14: 711–724.

de Groot, Mirjam. 2010. *Humans and Nature: Public Visions on Their Interrelationship.* Ph.D. Thesis, Radboud University, Nijmegen, The Netherlands. https://research.wur.nl/en/publications/humans-and-nature-public-visions-on -their-interrelationship

de Groot, Mirjam, Martin Drenthen, and Wouter T. de Groot. 2011. "Public visions of the human/nature relationship and their implications for environmental ethics." *Environmental Ethics* 33 (1): 25–44.

De Luccia, Thiago Paes de Barros. 2012. "*Mimosa pudica, Dionaea muscipula* and anesthetics." *Plant Signaling & Behavior* 7 (9): 1163–1167.

Descartes, René. 1976. "Animals are machines." In *Animal Rights and Human Obligations*, edited by Tom Regan and Peter Singer, 60–66. Englewood Cliifs: Prentice-Hall, Inc. Originally published in E.S. Haldane and G.R.T. Ross, trans.,

Discourse on Method in Philosophical Works of Descartes (London: Cambridge University Press), vol. I; and Anthony Kenny, trans. and ed. *Descartes: Philosophical Letters* (Oxford: Oxford University Press, 1970).

Descola, P. 1996. "Constructing natures: Symbolic ecology and social practice." In *Nature and Society: Anthropological Perspectives* (P. Descola and G. Pálsson, Eds.), pp. 82–102. London and New York: Routledge.

Devall, Bill, and William Devall. 1982. "Ecological consciousness and ecological resisting: Guidelines for comprehension and research." *Humboldt Journal of Social Relations* 9 (2): 177–196.

Devall, Bill. 2001. "The Deep, long-range ecology movement 1960–2000 – A review." *Ethics and the Environment* 6 (1): 18–41.

de Waal, F.B.M. 2007/2008. "Do animals feel empathy?" *Scientific American* 18 (6): 28–35.

Dhammika, Ven S. 1994. *The Edicts of King Asoka.* https://www.accesstoinsight.org/lib/authors/dhammika/wheel386.html [Accessed 30 September 2021].

Dhasmana, M.M. 1979. *The Ramos of Arunachal.* New Delhi: Concept Publishing Company.

Di Capua, Giuseppe, and Silvia Peppoloni. 2019. *Defining Geoethics.* Website of the IAPG - International Association for Promoting Geoethics. https://www.geoethics.org/definition [Accessed 25 July 2023].

Di Capua, G., S. Peppoloni, and P.T. Bobrowsky. 2017. "The Cape Town statement on geoethics." *Annals of Geophysics* 60 (Fast Rack 7): 2017. https://doi.org/10.4401/ ag-7553

Di Capua, G., P.T. Bobrowsky, S.W. Kieffer, and C. Palinkas. 2021. "Introduction: Geoethics goes beyond the geoscience profession." In *Geoethics: Status and Future Perspectives*, edited by G. Di Capua, P.T. Bobrowsky, S.W. Kieffer, and C. Palinkas, 1–11. London: Geological Society, Special Publications 508.

Dickinson, Emily. 2004. *Poems: Three Series, Complete.* Project Gutenberg's Poems: Three Series, Complete, by Emily Dickinson [EBook #12242] https://www.gutenberg.org/cache/epub/12242/pg12242.txt [Accessed 25 October 2023].

Dickerson, Adam. 2020. "Ecocentrism, economics and commensurability." *The Ecological Citizen* 3 (Suppl B): 5–11.

Dirks, Nicholas. 2021. "The ethics of sending humans to Mars." *Scientific American*, August 10. https://www.scientificamerican.com/article/the-ethics-of-sending-humans-to-mars/ [Accessed 25 July 2023].

Drengson, Alan, B. Devall, and M.A. Schroll. 2011. "The deep ecology movement: Origins, development and future prospects (toward a transpersonal ecosophy)." *International Journal of Transpersonal Studies* 30 (1–2): 101–117. http://doi.org/10.24972/ijts.2011.30.1-2.101

Duddy, Cathal. 2022a. *St. Francis – A Guide for Nature Lovers & Ecologists.* Praying Nature with St. Francis of Assisi. http://www.praying-nature.com/site_pages.php?section=Guide+for+Nature+Lovers

Duddy, Cathal. 2022b. *The Canticle of the Creatures by St. Francis of Assisi, 1225 AD.* Praying Nature with St. Francis of Assisi. http://www.praying-nature.com/site_pages.php?section=St.+Francis+of+Assisi&category_ref=1

Duddy, Cathal. 2022c. *An Extract from the Letter to the Faithful by St. Francis of Assisi c. 1220 AD.* Praying Nature with St. Francis of Assisi. http://www.praying-nature.com/site_pages.php?section=St.+Francis+of+Assisi&category_ref=19

Dunlap, Riley E. 2008. "The new environmental paradigm scale: From marginality to worldwide use." *The Journal of Environmental Education* 40 (1): 3–18.

Dunlap, Riley E., and Kent D. Van Liere. 1978. "The 'new environmental paradigm": A proposed measuring instrument and preliminary results." *The Journal of Environmental Education* 9 (4), 10–19.

Dunlap, Riley E., Kent D. Van Liere, Angela G. Mertig, and Robert Emmet Jones. 2000. "Measuring endorsement of the new ecological paradigm: A revised NEP scale." *Journal of Social Issues* 56: 425–442.

Dunlop, Rebecca, and Peter Laming. 2005. "Mechanoreceptive and nociceptive responses in the central nervous system of goldfish (*Carassius auratus*) and trout (*Onchorhynchus mykiss*)." *The Journal of Pain* 6 (9): 561–568.

Dunlop, Rebecca, Sarah Millsopp, and Peter Laming. 2006. "Avoidance learning in goldfish (*Carassius auratus*) and trout (*Oncorhynchus mykiss*) and implications for pain perception." *Applied Animal Behaviour Science* 97: 255–271.

Dutta, B.B. 1986. "Tradition vs. modernity: Problems of forestry planning in the Khasi Hills of Meghalaya." In *Forestry Development in North-East India*, edited by M. Das Gupta, A.K. Gangopadhyay, T. Bhattacharya, and M. Chakrabarti, 94–109. New Delhi: Omsons Publications.

Earhart, H.B. 1970. "The ideal of nature in Japanese religion and its possible significance for environmental concerns." *Contemporary Religions in Japan* 11 (1/2): 1–26.

Eco Jurisprudence Monitor. 2023. *Grant Township US Home Rule Charter*. Eco Jurisprudence Monitor. https://ecojurisprudence.org/initiatives/grant-township -home-rule-charter/ [Accessed 15 July 2023].

Edelman, David B., and Anil K. Seth. 2009. "Animal consciousness: A synthetic approach." *Trends in Neurosciences* 32 (9): 476–484.

Ehrlich, Paul, and Brian Walker. 1998. "Rivets and redundancy." *Bioscience* 48 (5): 387. https://doi.org/10.2307/1313377

Elbein, Saul. 2020. "In tiny Estonia, a fraught debate: What are forests for?" *National Geographic*. https://www.nationalgeographic.com/science/article/ estonia-holy-forests-threatened-by-industrial-tree-farming [Accessed 29 July 2022].

Elhakeem, Ali, Dimitrije Markovic, Anders Broberg, Niels P.R. Anten, and Velemir Ninkovic. 2018. "Aboveground mechanical stimuli affect belowground plant-plant communication." *PLoS One* 13 (5): e0195646. https://doi.org/10.1371/ journal.pone.0195646

Elwin, Verrier. 1949. *Myths of Middle India*. Madras: Oxford University Press.

Elwin, Verrier. 1954. *Tribal Myths of Orissa: Specimens of the Oral Literature of Middle India*. Oxford University Press.

Elwin, Verrier. 1958. *Myths of the North East Frontier of India*. Shillong: North-East Frontier Agency. https://archive.org/stream/in.ernet.dli.2015.104076/2015 .104076.Myths-Of-The-North-East-Frontier-Of-India_djvu.txt

Emerson, Ralph Waldo. 1849. *Nature*. Boston and Cambridge: James Munroe and Company. The Project Gutenberg EBook of Nature, by Ralph Waldo Emerson [EBook #29433], 2009.

Emerson, Ralph Waldo. 1867–1911. *Poems*. Household Edition. The Project Gutenberg EBook of Poems, by Ralph Waldo Emerson [EBook # 12843], 2004. https://www.gutenberg.org/ebooks/12843

Eurasiatic.eu. 2017. "Ancient Vasconic tree worship." http://eurasiatik.eu/2017/01 /27/ancient-vasconic-tree-worship/ [Accessed 03 August 2022].

Express News Service. 2017. "SC stays Uttarakhand HC order on Ganga, Yamuna living entity status." *The Indian Express*, July 8. https://indianexpress.com/ article/india/sc-stays-uttarakhand-hc-order-on-ganga-yamuna-living-entity -status-4740884/ [Accessed 16 July 2023].

Fabbro, Franco, Salvatore M. Aglioti, Massimo Bergamasco, Andrea Clarici, and Jaak Panksepp. 2015. "Evolutionary aspects of self- and world consciousness in vertebrates." *Frontiers in Human Neuroscience* 9: 157. https://doi.org/10.3389/ fnhum.2015.00157

Fan, Ruiping. 2005. "A reconstructionist Confucian account of environmentalism: Toward a human sagely dominion over nature." *Journal of Chinese Philosophy* 32 (1): 105–122.

Fox, Michael W. 1989. "The 'values' of sentient beings." *Between the Species* (Summer 1989): 158–159.

Frankena, W.K. 1973. *Ethics*, Second Edition. Prentice –Hall Foundations of Philosophy Series. Englewood Cliffs: Prentice-Hall, Inc.

Frazer, James George. 1890. *The Golden Bough: A Study of Magic and Religion*. London: Macmillan & Co. Temple of Earth Publishers. https://www.templeofearth.com/books/goldenbough.pdf [Accessed 22 November 2022].

Fromm, E. 1973. *The Anatomy of Human Destructiveness*. New York: Holt.

Furtak, Rick Anthony. 2019. "Henry David Thoreau." The Stanford Encyclopedia of Philosophy (Fall 2019 Edition), edited by Edward N. Zalta. https://plato.stanford.edu/archives/fall2019/entries/thoreau/ [Accessed 25 August 2023].

Gada, Mohd Yaseen. 2014. "Environmental ethics in Islam: Principles and perspectives." *World Journal of Islamic History and Civilization* 4 (4) 130–138.

Gadgil, M. 1995. "Prudence and profligacy: A human ecological perspective." In *The Economics and Ecology of Biodiversity Decline: The Forces Driving Global Change*, edited by T.M. Swanson, 99–110. Cambridge: Cambridge University Press.

Gadgil, M. and R. Guha. 1992. *This Fissured Land*. New Delhi: Oxford University Press.

Gagliano, Monica. 2017. "The mind of plants: Thinking the unthinkable." *Communicative & Integrative Biology* 10 (2): e1288333 (4 pages). http://doi.org/10.1080/19420889.2017.1288333

Gagliano, Monica, Vladyslav V. Vyazovskiy, Alexander A. Borbély, Martial Depczynski, and Ben Radford. 2020. "Comment on 'lack of evidence for associative learning in pea plants'." *eLife* 9: e61141. https://doi.org/10.7554/eLife.6114

Gagliano, Monica, Vladyslav V. Vyazovskiy, Alexander A. Borbély, Mavra Grimonprez, and Martial Depczynski. 2016. "Learning by association in plants." *Scientific Reports* 6 (38427): 1–9. https://doi.org/10.1038/srep38427

Gandhi, M.K. 1999. *The Collected Works of Mahatma Gandhi* (Electronic Book), 30–33, 36. New Delhi: Publications Division, Government of India.

Gell, Alfred. 1998. *Art and Agency: An Anthropological Theory*. Oxford: Clarendon Press.

Gerhardt, Christine. 2014. *A Place for Humility: Whitman, Dickinson, and the Natural World*. Iowa City: University of Iowa Press.

Gómez, Felipe. 2011. "Terrestrial analog." In *Encyclopedia of Astrobiology*, edited by M. Gargaud et al., 1654–1658. Berlin, Heidelberg: Springer. https://doi.org/10.1007/978-3-642-11274-4_1606 [Accessed 26 July 2023].

Goodpaster, Kenneth E. 1978. "On being morally considerable." *Journal of Philosophy* 75: 308–325.

Goralnik, L. and M.P. Nelson. 2012. "Anthropocentrism." In *Encyclopedia of Applied Ethics*, Second Edition, Volume 1, edited by Ruth Chadwick, 145–155. San Diego: Academic Press.

Grant, John. 2021. "The Moon's top layer alone has enough oxygen to sustain 8 billion people for 100,000 years." *The Conversation* https://theconversation.com/the-moons-top-layer-alone-has-enough-oxygen-to-sustain-8-billion-people-for-100-000-years-170013 [Accessed 18 July 2023].

Gray, Glenn J. 1957. "Heidegger's course: From human existence to nature." *The Journal of Philosophy* 54 (8): 197–207.

Green, Karen. 1996. "Two distinctions in environmental goodness." *Environmental Values* 5: 31–46.

Grémiaux, Alexandre, Ken Yokawa, Stefano Mancuso, and František Baluška. 2014. "Plant anesthesia supports similarities between animals and plants." *Plant Signaling & Behavior* 9 (1): e27886. https://doi.org/10.4161/psb.27886

Griffith, R.T.H., trans. 1889. *The Hymns of the Rigveda*, Second Edition, Kotagiri (Nilgiri) 1896. http://www.sanskritweb.net/rigveda/griffith.pdf [Accessed 30 September 2021].

Gronstal, Aaron. 2018. "Putting the ethics into planetary protection." *Astrobiology at NASA: Life in the Universe*. https://astrobiology.nasa.gov/news/putting-the-ethics-into-planetary-protection/ [Accessed 23 July 2023].

Gugliotta, G. 2008. "The great human migration." *History: Smithsonian Magazine.* https://www.smithsonianmag.com/history/the-great-human-migration-13561/ [Accessed 30 September 2021].

Gupta, Abhik. 2004. "Cultural diversity-biodiversity-traditional medicine linkages in India: An ecoethical impact analysis." In *Challenges for Bioethics from Asia*, edited by Darryl R.J. Macer, 204–210. Christchurch and Tsukuba Science City: Eubios Ethics Institute.

Gupta, Abhik. 2008. "From biosphere to technosphere to biotechnosphere: The Indian scenario in an eco-ethical perspective." In *Asia-Pacific Perspectives on Environmental Ethics*, edited by D.R.J. Macer, 22–29. Bangkok: UNESCO Bangkok.

Gupta, Abhik. 2013. "Altruism in Indian religions: Embracing the biosphere." In *Altruism in Cross-Cultural Perspective*, edited by Douglas A. Vakoch, 101–112. New York: Springer.

Gupta, Abhik. 2016. "Ecocentric thoughts in the writings of Rabindranath Tagore: An illustrative study." In *The Idea of Surplus: Tagore and Contemporary Human Sciences*, edited by Mrinal Miri, 235–248. New Delhi, London, and New York: Routledge & Indian Council of Social Science Research, North Eastern Regional Centre, Shillong, India.

Gupta, Abhik. 2022. "Anxiety in isolation: Anointing with ecocentrism." In *Eco-Anxiety and Planetary Hope: Experiencing the Twin Disasters of COVID-19 and Climate Change*, edited by Douglas A. Vakoch and Sam Mickey, 119–126. Cham: Springer.

Gupta, Abhik. 2023. "Bengali literature and ecofeminism." In *The Routledge Handbook of Ecofeminism and Literature*, edited by Douglas A. Vakoch, 88–100. New York and London: Routledge.

Gupta, Abhik, and Kamalesh Guha. 2002. "Tradition and conservation in Northeastern India: An ethical analysis." *Eubios Journal of Asian and International Bioethics* 12: 15–18.

Gupta, Abhik, and Suchandra Ghosh. 2003. "Wildlife conservation in the Maurya period." *Journal of Ancient Indian History* (Vol. XXI, 2000–2002) (Kalyan Kumar Dasgupta Memorial Volume): 142–152. Kolkata: University of Calcutta.

Hall, Mathew. 2020. "Lunar gold rush: Can Moon mining ever take off?" *Mining Technology* https://www.mining-technology.com/analysis/moon-mining-what-would-it-take/ [Accessed 18 July 2023].

Halper, Phil, Kenneth Williford, David Rudrauf, and Perry N. Fuchs. 2021. "Against neo-cartesianism: Neurofunctional resilience and animal pain." *Philosophical Psychology* 34 (4): 474–501.

Hamilton, Clive, and Jacques Grinevald. 2015. "Was the Anthropocene anticipated?" *The Anthropocene Review* 2 (1): 59–72. https://doi.org/10.1177/2053019614567155

Handmer, Casey J. 2022. "How to terraform Mars for $10b in 10 years." *Casey Handmer's Blog, Wordpress.com.* https://caseyhandmer.wordpress.com/2022/07/12/how-to-terraform-mars-for-10b-in-10-years/ [Accessed 23 July 2023].

Hardie, W.F.R. 1964–1965. "Aristotle's doctrine that virtue is a 'mean'." *Proceedings of the Aristotelian Society, New Series* 65 (1964–1965): 183–204.

Harold, J. 2005. "Between intrinsic and extrinsic value." *Journal of Social Philosophy* 36 (1): 85–105.

Hart, Amalyah. 2023. "Mining the moon: Do we have the right?" *Cosmos*, February 3. https://cosmosmagazine.com/space/mining-the-moon/ [Accessed 19 July 2023].

Hartman, Steven. 2007. "The rise of American ecoliterature." *American Studies in Scandinavia* 39 (2): 41–58.

Hawcroft, Lucy J., and Taciano L. Milfont. 2010. "The use (and abuse) of the new environmental paradigm scale over the last 30 years: A meta-analysis." *Journal of Environmental Psychology* 30 (2): 143–158.

Hayward, Tim. 1997. "Anthropocentrism: A misunderstood problem." *Environmental Values* 6 (1): 49–63.

Heidegger, Martin. 1977. *The Question Concerning Technology and Other Essays.* Translated by William Lovitt. New York and London: Garland Publishing, Inc.

Henrich, J. and R. McElreath. 2002. "Are peasants risk-averse decision makers?" *Current Anthropology* 43: 172–181.

Hirai, N. 1960. "The principles of shrine Shinto." *Japanese Journal of Religious Studies* 1 (1): 39–54. https://nirc.nanzan-u.ac.jp/nfile/3260 [Accessed 30 September 2021].

Hoefle, Scott William. 2009. "Enchanted (and disenchanted) Amazonia: Environmental ethics and cultural identity in Northern Brazil." *Ethics, Place and Environment* 12 (1): 107–130.

Holtom, D.C. 1940. "The meaning of kami." Chapter II. Interpretations by Japanese writers. *Monumenta Nipponica* 3 (2): 392–413.

Holy-Lucsaz, Magdalena. 2015. "Heidegger's support for Deep Ecology reexamined once again: Ontological egalitarianism, or farewell to the great chain of being." *Ethics and the Environment* 20 (1): 45–66.

Howe, Lawrence W. 1993. "Heidegger's discussion of 'The Thing': A theme for deep ecology." *Between the Species* (Spring): 93–97.

Hume, David. 1777. *An Enquiry Concerning the Principles of Morals.* A 1912 Reprint of the Edition of 1777. The Project Gutenberg EBook of *An Enquiry Concerning the Principles of Morals, by David Hume*, 2010 [EBook # 4320]. https://www.gutenberg.org/files/4320/4320-h/4320-h.htm [Accessed 10 August 2023].

Hutchinson, G. Evelyn. 1970. "The biosphere." *Scientific American* 223 (3): 44–53.

Indian Culture. 2022. *Indigo Revolt in Bengal.* Ministry of Culture, Government of India, in partnership with Indian Institute of Technology, Bombay, and Indira Gandhi National Open University, New Delhi. https://indianculture.gov.in/stories/indigo-revolt-bengal [Accessed 14 July 2022].

Ingold, T. 2006. "Rethinking the animate, re-animating thought." *Ethnos* 71 (1): 9–20. https://doi.org/10.1080/0014184060060311

Insoll, T. 2011. "Animism and totemism." In *Oxford Handbook of the Archaeology of Ritual and Religion*, edited by T. Insoll, Chapter 62, 1004–1016. Oxford: Oxford University Press.

International Union for Conservation of Nature (IUCN). 2022. *The IUCN Red List of Threatened Species.* Version 2021–3. https://www.iucnredlist.org/

International Union for Conservation of Nature (IUCN). 1980. *World Conservation Strategy: Living Resource Conservation for Sustainable Development.* International Union for Conservation of Nature (IUCN), with the advice, cooperation and financial assistance of the United Nations Environment

Programme (UNEP), and the World Wildlife Fund (WWF), in collaboration with the Food and Agriculture Organization of the United Nations (FAO), and the United Nations Educational, Scientific and Cultural Organization (UNESCO). https://portals.iucn.org/library/efiles/documents/wcs-004.pdf [Accessed 05 July 2023].

IPBES. 2019. "Summary for policymakers of the global assessment report on biodiversity and ecosystem services of the Intergovernmental Science-Policy Platform on Biodiversity and Ecosystem Services." In edited by S. Díaz, J. Settele, E. Brondízio, H.T. Ngo, M. Guèze, J. Agard, et al. Bonn: IPBES Secretariat. https://zenodo.org/records/3553579 [Accessed 19 October 2023].

Jamieson, Paul F. 1958. "Emerson in the Adirondacks." *New York History* 39 (3): 215–237.

Janowski, M. 2020. "Stones alive! An exploration of the relationship between humans and stone in southeast Asia." *Bijdragen Tot De Taal-, Land – En Volkenkunde* 176: 105–146.

Johnson, Robert, and Adam Cureton. 2022. "Kant's moral philosophy." In *The Stanford Encyclopedia of Philosophy*, edited by Edward N. Zalta (Spring 2022 Edition). https://plato.stanford.edu/archives/spr2022/entries/kant-moral/ [Accessed 10 May 2022].

Kauffman, Craig M., and Pamela L. Martin. 2023. "Five native groups have established rights of nature." *Barn Raiser*, April 17. https://barnraisingmedia.com/rights-of-nature-native-ho-chunk-ponca-chippewa-ojibwe-yurok/ [Accessed 14 July 2023].

Keeter, Bill, ed. 2021. *Space Colonization*. National Aeronautics and Space Administration (NASA). https://www.nasa.gov/centers/hq/library/find/bibliographies/space_colonization [Accessed 17 July 2023].

Keith, A.B., trans. 1914. *The Veda of the Black Yajus School entitled Taittiriya Sanhita*. The Harvard Oriental Series, Volume Nineteen (edited by C.R. Lanman). Cambridge: The Harvard University Press. https://ia800304.us.archive.org/32/items/vedablackyajuss00keitgoog/vedablackyajuss00keitgoog.pdf [Accessed 30 September 2021].

Keller, David. 1997. "Gleaning lessons from deep ecology." *Ethics and the Environment* 2 (2): 139–148.

Keller, David R. 2009. "Deep ecology." In *Encyclopedia of Environmental Ethics and Philosophy*, Volume 1: Abbey to Israel, edited by J. Baird Callicott, Robert Frodeman (Editors in Chief), 206–211. Detroit: Macmillan Reference USA, A part of Gale, Cengage Learning.

Kellert, Stephen R. 1996. *The Value of Life*. Washington, DC: Island Press.

Kellert, Stephen R., and Edward O. Wilson. 1993. *The Biophilia Hypothesis*. Washington, DC: Island Press.

Kerstein, Samuel. 2019. "Treating persons as means." In *The Stanford Encyclopedia of Philosophy*, edited by Edward N. Zalta (Summer 2019 Edition), https://plato.stanford.edu/archives/sum2019/entries/persons-means/ [Accessed 13 July 2022].

Key, Brian. 2016. "Why fish do not feel pain." *Animal Sentience* 3 (1). https://doi.org/10.51291/2377-7478.1011

Kidner, David W. 2014. "Why 'anthropocentrism' is not anthropocentric." *Dialectical Anthropology* 38: 465–480.

Killingsworth, M.J. 2010. "'As if the beasts spoke': The Animal/Animist/Animated Walt Whitman." *Walt Whitman Quarterly Review* 28 (1/2): 19–35. https://doi.org/10.13008/2153-3695.1949

Knecht, T. 2004. "What is Machiavellian intelligence?" *The Neurologist* 75: 1–5. https://doi.org/10.1007/s00115-003-1543-0

Kopnina, Helen, Haydn Washington, Bron Taylor, and John J. Piccolo, 2018. "Anthropocentrism: More than just a misunderstood problem." *Journal of Agricultural Environmental Ethics*. https://doi.org/10.1007/s10806-018-9711-1

Kopnina, Helen. 2020. "Anthropocentrism and post-humanism." In *The International Encyclopedia of Anthropology*, edited by Hilary Callan, 1–8. Hoboken: John Wiley & Sons, Ltd.

Korsgaard, Christine M. 1983. "Two distinctions in goodness." *The Philosophical Review* XCII (2): 169–195.

Krause, Jens, Graeme D. Ruxton, and Stefan Krause. 2009. "Swarm intelligence in animals and humans." *Trends in Ecology and Evolution* 25 (1): 28–34. https://doi.org/10.1016/j.tree.2009.06.016

Kriebel, David, Joel Tickner, Paul Epstein, John Lemons, Richard Levins, Edward L. Loechler, Margaret Quinn, Ruthann Rudel, Ted Schettler, and Michael Stoto. 2001. "The precautionary principle in environmental science." *Environmental Health Perspectives* 109 (9): 871–876.

Kumar, Victor. 2016. "Myth, magic, and mind in *The Golden Bough*." *HAU: Journal of Ethnographic Theory* 6 (2): 233–254.

Kumarappa, J.C. 1957. *Economy of Permanence: [A Quest for a Social Order Based on Non-Violence]*, Third Edition. Varanasi: Sarva Seva Sangh Prakashan.

La Fleur, W.R. 1973. "Saigyō and the Buddhist value of nature." *History of Religions* 13 (2): 93–128.

Lai, K. 2015. "Daoism and confucianism." In *Dao Companion to Daoist Philosophy*, Volume 6, edited by X. Liu, 489–511. Dordrecht: Dao Companions to Chinese Philosophy, Springer.

Lai, K.L. 2003. "Conceptual foundations for environmental ethics: A Daoist perspective." *Environmental Ethics* 25: 247–266.

Lakoff, George, and Mark Johnson. 1980/2003. *The Metaphors We Live By*. London: The University of Chicago Press.

Langford, D.J., Crager, S.E., Shehzad, Z. et al. 2006. "Social modulation of pain as evidence for empathy in mice." *Science* 312: 1967–1970.

Leopold, Aldo. 1949. *A Sand County Almanac and Sketches Here and There*. New York: Oxford University Press. American Museum of Natural History Special Members' Edition.

Le Roy, Edouard. 1999. "The origins of humanity and the evolution of mind." In *The Biosphere and Noosphere Reader*, edited by Paul R. Samson and David Pitt, 60–70. Taylor and Francis e-Library Edition, 2002. London and New York: Routledge.

Letenyei, Danielle. 2023. "If we nuke mars, would it really become an earth-like planet?" *Green Matters*, June 1. https://www.greenmatters.com/technology/nuke-mars [Accessed 22 July 2023].

Li, Alex S. 2023. "Touring outer space: The past, present, and future of space tourism." *Cleveland State Law Review* 71 (3): 743. https://engagedscholarship.csuohio.edu/clevstlrev/vol71/iss3/8 [Accessed 20 July 2023]

Light, Andrew. 2002. "Contemporary environmental ethics from metaethics to public philosophy." *Metaphilosophy* 33 (4): 426–449.

Live Law News Network. 2017. "A first in India: Uttarakhand HC declares Ganga, Yamuna Rivers as living legal entities." https://www.livelaw.in/first-india-uttarakhand-hc-declares-ganga-yamuna-rivers-living-legal-entities/?infinitescroll=1

Loy, Ignacio, Susana Carnero-Sierra, Félix Acebes, Judit Muñiz-Moreno, Clara Muñiz-Diez, and José-Carlos Sánchez-González. 2021. "Where association ends: A review of associative learning in invertebrates, plants and Protista, and a reflection on its limits." *Journal of Experimental Psychology: Animal Learning and Cognition* 47 (3): 234–251.

Lundmark, Carina. 2007. "The new ecological paradigm revisited: anchoring the NEP scale in environmental ethics." *Environmental Education Research* 13 (3): 329-347. DOI: 10.1080/13504620701430448

Lurgio, Jeremy. 2019. "Saving the Whanganui: Can personhood rescue a river?" *The Guardian*, November 29. https://www.theguardian.com/world/2019/nov/30/saving-the-whanganui-can-personhood-rescue-a-river [Accessed 13 July 2023].

Majumdar, G.P. 1927. *Vanaspati – Plants and Plant-Life as in Indian Treatises and Traditions* [Griffith Memorial Prize Essay for 1925]. Kolkata: University of Calcutta.

Malhotra, Kailash C., Yogesh Gokhale, Sudipto Chatterjee, and Sanjeev Srivastava. 2001. *Cultural and Ecological Dimensions of Sacred Groves in India*. New Delhi: Indian National Science Academy, & Bhopal: Indira Gandhi Rashtriya Manav Sangrahalaya.

Mallatt, Jon, Lincoln Taiz, Andreas Draguhn, Michael R. Blatt, and David G. Robinson. 2021. "Integrated information theory does not make plant consciousness more convincing." *Biochemical and Biophysical Research Communications* 564: 166–169.

Mamo, Dwayne, ed. 2022. *The Indigenous World 2022*, Thirty Sixth Edition. Copenhagen: International Work Group for Indigenous Affairs (IWGIA). https://www.iwgia.org/doclink/iwgia-book-the-indigenous-world-2022-eng/eyJ0eXAiOiJKV1QiLCJhbGciOiJIUzI1NiJ9.eyJzdWIiOiJpd2dpYS1ib29rLXRRoZS1pbmRpZ2Vub3VzLXdvcmxkLTIwMjItZW5nIiwiaWaWF0IjoxNjUxMTM5NTg1LCJleHAiOjE2NTEyMjU5ODV9.jRnv3PeantfRZtJg4jph8xdshK5Mh25Z3hlcPs9As_U [Accessed 25 September 2022].

Manoli, Constantinos C., Bruce Johnson, Sanlyn Buxner, and Franz Bogner. 2019. "Measuring environmental perceptions grounded on different theoretical models: The 2-major environmental values (2-MEV) model in comparison with the new ecological paradigm (NEP) scale." *Sustainability* 11: 1286. https://doi.org/10.3390/su11051286

Margil, Mari. 2017. "Court decisions advance legal rights of nature globally." In *Rights of Nature & Mother Earth: Rights-Based Law for Systemic Change*, edited by Shannon Briggs, Tom B.K. Goldtooth, and Osprey Orielle Lake, 27–28. Oakland, CA: Movement Rights, Women's Earth & Climate Action Network, Indigenous Environmental Network.

Markel, Kasey. 2020a. "Lack of evidence for associative learning in pea plants." *eLife* 9: e57614. https://doi.org/10.7554

Markel, Kasey. 2020b. "Response to comment on 'lack of evidence for associative learning in pea plants'." *eLife* 9: e61689. https://doi.org/10.7554/eLife.61689

Marino, Lori. 2010. "Sentience." In *Encyclopedia of Animal Behavior*, Volume 3, edited by M. Breed and J. Moore, 132–138. Oxford: Academic Press.

Marino, Lori. 2017. "Thinking chickens: A review of cognition, emotion and behavior in the domestic chicken." *Animal Cognition* 20: 127–147.

Martin, Adam. 2023. "Bolivia's "Law of Mother Earth" – What is it and is it working?" *Wildlife Calendar – Wildlife & Welfare*. https://www.wildlifeandwelfare.org/news-blog/bolivias-law-of-mother-earth

Martin, Neil. 2022. "8 things you never knew about mining on Mars, the Moon …. and even asteroids." *UNSW Newsroom*, May 12. https://newsroom.unsw.edu.au/news/science-tech/8-things-you-never-knew-about-mining-mars-moon-and-even-asteroids [Accessed 19 July 2023].

Martinez-Frias, Jesus, Gerda Horneck, Rosa De La Torre Noetzel, and Fernando Rull. 2010. "A geoethical approach to the geological and astrobiological exploration and research of the Moon and Mars." In *38th COSPAR Scientific Assembly 2010 – Panels (P): Protecting the Lunar and Martian Environments for Scientific Research (PEX1)*. https://www.cospar-assembly.org/abstractcd/ OLD/COSPAR-10/abstracts/data/pdf/abstracts/PEX1-0007-10.pdf [Accessed 25 July 2023].

Martinez-Frias, J., E.M. Mederos, and R. Lunar. 2017. "The scientific and educational significance of geoparks as planetary analogues: The example of Lanzarote and Chinijo Islands UNESCO Global Geopark." *Episodes* 40 (4): 343–347. http://doi.org/10.18814/epiiugs/2017/v40i4/017035

Meents, Anja K., and Mithöfer, Axel. 2020. "Plant-plant communication: Is there a role for volatile damage-associated molecular patterns?" *Frontiers in Plant Science* 11: 583275. https://doi.org/10.3389/fpls.2020.583275

Meyer, John M. 1997. "Gifford Pinchot, John Muir, and the boundaries of politics in American thought." *Polity* 30 (2): 267–284.

Miller, J. 2003. "Daoism and nature." In *Nature Across Cultures. Science Across Cultures: The History of Non-Western Science*, Volume 4, edited by H. Selin, 393–409. Dordrecht: Springer. https://doi.org/10.1007/978-94-017-0149-5_20

Minorsky, Peter V. 2021. "American racism and the lost legacy of Sir Jagadis Chandra Bose, the father of plant neurobiology." *Plant Signaling & Behavior* 16 (1): 1818030, https://doi.org/10.1080/15592324.2020.1818030

Montani, Guido. 2007. "The ecocentric approach to sustainable development. Ecology, economics and politics." *The Federalist Political Revue* XLIX (1): 12.

Muir, John. 1901. *Our National Parks*. Boston New York, and Cambridge: Houghton, Mifflin and Company, The Riverdale Press. Digitized by the Internet Archive in 2007. https://ia800906.us.archive.org/13/items/nationalparksour00m uirrich/nationalparksour00muirrich.pdf [Accessed 20 September 2022].

Mukherjee, B.N. 1984. *Studies in Aramaic Edicts of Aśoka*. Calcutta: Indian Museum.

Mukherjee, B.N. 2000. *The Character of the Maurya Empire*. Calcutta: Pilgrim.

Munévar, Gonzalo. 2014. "Damasio, self and consciousness." *Philosophia Scientae* 18 (3): 191–201. https://doi.org/10.4000/philosophiascientiae.1015

Muñoz, Lorna. 2023. "Bolivia's mother earth laws: Is the ecocentric legislation misleading?" *ReVista: Harvard Review of Latin America*. https://revista .drclas.harvard.edu/bolivias-mother-earth-laws-is-the-ecocentric-legislation -misleading/ [Accessed 24 July 2023].

Munro, Daniel. 2022. "If humanity is to succeed in space, our ethics must evolve." Waterloo: Center for International Governance Innovation. https://www .cigionline.org/articles/if-humanity-is-to-succeed-in-space-our-ethics-must -evolve/ [Accessed 21 July 2023].

Muraca, Barbara. 2011. "The map of moral significance: A new axiological matrix for environmental ethics." *Environmental Values* 20: 375–396.

Musschenga, A.W. 2002. "Naturalness: Beyond animal welfare." *Journal of Agricultural and Environmental Ethics* 15: 171–186.

Naess, Arne. 1973. "The shallow and the deep, long-range ecology movement: A summary." *Inquiry: An Interdisciplinary Journal of Philosophy and the Social Sciences* 16: 95–100.

Naess, Arne. 1987. "Self-realization: An ecological approach to being in the world." *Trumpeter* 4 (3): 35–42.

Naess, Arne. 1989. *Ecology, Community and Lifestyle: Outline of an Ecosophy*. Translated and Revised by David Rothenberg. Cambridge: Cambridge University Press.

Naess, Arne. 1997. "Heidegger, postmodern theory and deep ecology." *Trumpeter* 14 (4). http://www.icaap.org/iuicode?6.14.4.5

National Assembly of Panama. 2022. *Law 287: 24th of February 2022, Which Recognizes the Rights of Nature and the Duties of the State in Relation to*

Said Rights. Official Digital Gazette. February 24. https://static1.squarespace
.com/static/5e3f36df772e5208fa96513c/t/623a3e7d89eb3a28d04d45a3
/1647984253637/2022+PANAMA+RIGHTS+OF+NATURE+LAW+IN
+ENGLISH.pdf [Accessed 15 July 2023].

National Agency for Space Administration (NASA). 2020. *The Artemis Accords: Principles for Cooperation in the Civil Exploration and Use of the Moon, Mars, Comets, and Asteroids for Peaceful Purposes.* https://www.nasa.gov/specials/
artemis-accords/img/Artemis-Accords-signed-13Oct2020.pdf [Accessed 19 July 2023].

Newman, Christopher J. 2015. "Seeking tranquility: Embedding sustainability in lunar exploration policy." *Space Policy* 33: 29–37.

New Zealand Legislation. 2021. *Te Urewera Act 2014.* Parliamentary Counsel Office (*Te Tari Tohutohu Pāremata*) – Version as at 28 October 2021 (Public Act 2014 No 51; Date of assent 27 July 2014). https://www.legislation.govt.nz/act
/public/2014/0051/latest/whole.html#DLM6183705 [Accessed 12 July 2023].

New Zealand Legislation. 2022. *Te Awa Tupua (Whanganui River Claims Settlement) Act 2017.* Parliamentary Counsel Office (*Te Tari Tohutohu Pāremata*) – Version as at 30 November 2022 (Public Act 2017 No 7; Date of assent 20 March 2017). https://www.legislation.govt.nz/act/public/2017/0007/
latest/whole.html#DLM6830851 [Accessed 12 July 2023].

Ng, Yew-Kwang. 2016. "Could fish feel pain? A wider perspective." *Animal Sentience* 2016.019. https://doi.org/10.51291/2377-7478.1030

Nicholson, Wayne L., Andrew C. Schuerger, and Margaret S. Race. 2009. "Migrating microbes and planetary protection." *Trends in Microbiology* 17 (9): 389–392.

Nikitina, Nataliya. 2016. *Geoethics: Theory, Principles, Problems,* Second Edition, revised and supplemented. Moscow: M.: Geoinformmark, Ltd. https://www.icog
.es/iageth/files/Nikitina_Geoethics.pdf [Accessed 25 July 2023].

Nobutaka, I. 2000. *Perspective Toward Understanding the Concept of Kami.* Institute for Japanese Culture and Classics, Kokugakuin University. https://www2
.kokugakuin.ac.jp/ijcc/wp/cpjr/kami/intro.html [Accessed 30 September 2021].

Norton, B.G. 1984. "Environmental ethics and weak anthropocentrism." *Environmental Ethics* 6 (2): 131–148.

Nuyen, Anh Tuan. 2011. "Confucian role-based ethics and strong environmental ethics." *Environmental Values* 20 (4): 549–566.

Odum, E.P. 1971. *Fundamentals of Ecology,* Third Edition. Philadelphia: W.B. Saunders.

Oladipo, Gloria, and Agency. 2023. "Nasa aims to mine resources on moon in next decade." *The Guardian,* June 28. https://www.theguardian.com/science/2023/
jun/28/nasa-mining-moon-2032#:~:text=US [Accessed 19 July 2023].

Owe, Andrea. 2023. "Greening the Universe: The case for ecocentric space expansion." In *Reclaiming Space: Progressive and Multicultural Visions of Space Exploration,* edited by James S.J. Schwarz, Linda Billings, and Erica Nesvold, 425–336. Oxford: Oxford University Press. https://doi.org/10.1093/oso
/9780197604793.003.0027

Özdemir, İbrahim. 2019. "A bestowed trust: The perception of nature and animals in Islam." *The Ecological Citizen* 3 (1): 33–34.

Pachauri, Swasti, and Camille Parker. 2018. "The blue boom." *Down to Earth,* April 15. https://www.downtoearth.org.in/news/agriculture/the-blue-boom
-60049 [Accessed 14 July 2022].

Parise, André Geremia, Monica Gagliano, and Gustavo Maia Souza. 2020. "Extended cognition in plants: Is it possible?" *Plant Signaling & Behavior* 15 (2): e1710661 (5 pages). https://doi.org/10.1080/15592324.2019.1710661

Parks, Jake. 2019. "Moon village: Humanity's first step toward a lunar colony?" *Astronomy*, May 31. https://www.astronomy.com/observing/moon-village -humanitys-first-step-toward-a-lunar-colony/ [Accessed 19 July 2023].

Paton, G.W. *A Text-Book of Jurisprudence*. Third Edition, edited by David P. Derham. Oxford: At the Clarendon Press.

Pedersen, M.A. 2001. "Totemism, animism and North Asian indigenous ontologies." *Journal of Royal Anthropological Institute* 7: 411–427.

Pennsylvania Department of Community and Economic Development. 2020. *Home Rule in Pennsylvania*. Pennsylvania Governor's Center for Local Government Services. https://www.cityoflancasterpa.gov/wp-content/uploads/2023/02/ HomeRule_2020.pdf [Accessed 15 July 2023].

Peoples, H.C., P. Duda, and F.W. Marlowe. 2016. "Hunter-gatherers and the origins of religion." *Human Nature* 27: 261–282.

Peppoloni, S., and G. Di Capua. 2016. "Geoethics: Ethical, social and cultural values in geosciences research, practice and education." In *Geosciences for the Public Good and Global Development: Toward a Sustainable Future*, Volume 520, edited by G.R. Wessel, and J.K. Greenberg, 1–9. Geological Society of America. https://doi.org/10.1130/SPE520

Perez, Jason, ed. 2019. *Houghton Mars Project (HMP)*. https://www.nasa.gov/ analogs/hmp [Accessed 17 July 2023].

Piccolo, John J. 2017. "Intrinsic values in nature: Objective good or simply half of an unhelpful dichotomy?" *Journal for Nature Conservation* 37: 8–11.

Plurinational State of Bolivia. 2010. *Law of the Rights of Mother Earth (Ley de Derechos de la Madre Tierra)*. Law of Mother Earth – The Rights of Our Planet: A Vision from Bolivia. Alexandria: World Future Fund. http://www .worldfuturefund.org/projects/indicators/motherearthbolivia.html [Accessed 07 July 2023].

Plurinational State of Bolivia. 2012. *Framework Law of Mother Earth and Integral Development to Live Well (Ley Marco de la Madre Tierra Y Desarrollo Integral Para Vivir Bien)*. https://ecojurisprudence.org/wp-content/uploads/2022/02 /Bolivia_Law-No.-300-the-Framework-Law-of-Mother-Earth-and-Integral -Development-to-Live-Well_70.pdf. Derechos Reservados © 2012 www.gacetao ficialdebolivia.gob.bo [Accessed 10 July 2023].

Polunin, Nicholas, and Jacques Grinevald. 1999. "Vernadsky and biospheral ecology." In *The Biosphere and Noosphere Reader*, edited by Paul R. Samson and David Pitt, 94–27. Taylor and Francis e-Library Edition, 2002. London and New York: Routledge.

Pope Francis. 2015. *Encyclical Letter Laudato Si' of the Holy Father Francis: On Care for Our Common Home*. Rome: Vatican Press, 18 June 2015.

Poppick, Laura. 2017. "When humans begin colonizing other planets, who should be in charge?" Think Big: A Smithsonian magazine special report. https://www .smithsonianmag.com/science-nature/humans-begin-colonizing-other-planets -who-should-be-in-charge-180962331/ [Accessed 17 July 2023].

Prabhu, R.K. and U.R. Rao (Compiled and edited). 1967. *The Mind of Mahatma Gandhi: Encyclopedia of Gandhi's Thoughts*. Ahmedabad: Navajivan Mudranalaya.

ProGEO. 1991. *Declaration Digne-Les-Bains: Declaration of the Rights of the Memory of the Earth*. The European Association for the Conservation of the Geological Heritage. chrome-extension://efaidnbmnnnibpcajpcglclefindmkaj/ http://www.progeo.ngo/downloads/DIGNE_DECLARATION.pdf

Raffensperger, Carolyn, and Joel Tickner, eds. 1999. *Protecting Public Health and the Environment: Implementing the Precautionary Principle*. Washington, DC, Covelo: Island Press.

Rafferty, John P. 2023. *Anthropocene Epoch*. Encyclopaedia Britannica. https://www.britannica.com/science/Anthropocene-Epoch [Accessed 25 August 2023].

Rahman, Aminur Khosru. 2022. *Islamic Environmental Ethics: A Model for Shaping Muslim Attitudes in Helping to Promote Environmental Education, Awareness and Activism*. Masters Thesis, University of Wales, Trinity Saint David. https://repository.uwtsd.ac.uk/id/eprint/2032/1/Rahman%2C%20K%20%20%282022%29%20MRes%20Islamic%20environmental%20ethics.pdf [Accessed 25 August 2023].

Raj, Arpita. 2019. "Revisiting human and non-human relationship in *Santal* worldview." *Lokaratna* XII: 10–16.

Rajan, R. 1998. "Imperial environmentalism or environmental imperialism? European forestry, colonial foresters and the agendas of forest management in British India 1800–1900." In *Nature and the Orient: The Environmental History of South and Southeast Asia*, edited by R. Grove, V. Damodaran, and S. Sangwan, 324–371. New Delhi: Oxford University Press.

Randall, Roderick P. 2012. *A Global Compendium of Weeds*, Second Edition. Perth: Department of Agriculture and Food, Western Australia.

Rao, P. 1996. "Sacred groves and conservation." *WWF – India Quarterly* 7: 4–8.

Rea, Anne W., Munns Jr, Wayne R. 2017. "The value of nature: Economic, intrinsic or both?" *Integrated Environmental Assessment and Management* September 13 (5): 953-955. https://doi.org/10.1002/ieam.1924

Regan, Tom. 1983. *The Case for Animal Rights*. Berkeley, Los Angeles: University of California Press.

Republic of Ecuador. 2011. *Republic of Ecuador Constitution of 2008*. National Assembly, Legislative and Oversight Committee: Published in the Official Register, October 20, 2008. Georgetown University, Edmund A. Walsh School of Foreign Services, Center for Latin American Studies Program.: Political Database of the Americas. https://pdba.georgetown.edu/Constitutions/Ecuador/english08.html [Accessed 06 July 2023].

Rhi, B.Y. 1993. "The phenomenology and psychology of Korean shamanism." In *Contemporary Philosophy: A New Survey*, Volume 7, *Asian Philosophy*, edited by Guttorm Fløistad, 253–268. Dordrecht: Kluwer Academic Publishers. https://doi.org/10.1007/978-94-011-2510-9_14

Richardson, Eliza. 2022. *Eduard Suess*. Earth 520: Plate Tectonics and People. Penn State College of Earth and Mineral Sciences. https://www.e-education.psu.edu/earth520/node/1795 [Accessed 21 June 2023].

Richardson, William. 2003. *Heidegger – Through Phenomenology to Thought*. New York: Fordham University Press.

Richerson, P.J. and R. Boyd. 2001. "The evolution of subjective commitment to groups: A tribal instincts hypothesis." In *Evolution and the Capacity of Commitment*, edited by R.M. Nesse, 186–220. New York: Russell Sage.

Richerson, P.J., R.T. Boyd, and J. Henrich. 2003. "Cultural evolution of human cooperation." In *Genetic and Cultural Evolution of Cooperation*, edited by Peter Hammerstein, 357–388. Cambridge: The MIT Press.

Ritchie, Hannah, and Max Roser. 2021. *Forests and Deforestation*. OurWorldInData.org. https://ourworldindata.org/deforestation [Accessed 03 July 2023].

Ritter, Eva, and Dainis Dauksta. 2006. "Ancient values and contemporary interpretations of European forest culture – Reconsidering our understanding of sustainability in forestry." In *Small-scale Forestry and Rural Development: The Intersection of Ecosystems, Economics and Society*. Proceedings of IUFRO 3.08 Conference, pp. 424–432. Hosted by Galway-Mayo Institute of Technology, Galway. http://www.coford.ie/media/coford/content/publications/projectreports/small-scaleforestryconference/Ritter.pdf [Accessed 22 November 2022].

Robertson, Miranda. 2009. "Ockham's broom: A new series." *Journal of Biology* 8: 79.

Robinson, Shorty Jangala. 2022. "Ngapa Jukurppa (water dreaming)." https://www.kateowengallery.com/page/Water-Dreaming [Accessed 16 September 2022].

Rolston, Holmes III. 1985. "Duties to endangered species." *Bioscience* 35 (11): 718–726. The Biological Diversity Crisis.

Rolston, Holmes III. 1988. *Environmental Ethics: Duties to and Values in the Natural World*. Philadelphia: Temple University Press.

Rolston, Holmes III. 1994. "Value in nature and the nature of value." In *Philosophy and the Natural Environment*, Royal Institute of Philosophy Supplement 36, edited by Robin Attfield and Andrew Belsey, 13–30. Cambridge: Cambridge University Press.

Rorty, Richard. 1991. *Objectivity, Relativism and Truth*. Cambridge: Cambridge University Press.

Rose, J.D., R. Arlinghaus, S.J. Cooke, B.K. Diggles, W. Sawynok, E.D. Stevens, and C.D.L. Wynne. 2012. "Can fish really feel pain?" *Fish and Fisheries* https://doi.org/10.1111/faf.12010

Rummel, John D. 2017. "Some ethical considerations in space explorations." *Harvard Business School, Working Group on the Business and Economics of Space at Harvard Business School* (November 3–4, 2017). https://www.hbs.edu/faculty/Shared%20Documents/conferences/2017-business-and-economics-of-space/John%20Rummel.pdf [Accessed 21 July 2023].

Rutherford, A., A.B. Zwi, N.J. Grove, and A. Butchart. 2007. "Violence: A glossary." *Journal of Epidemiology and Community Health* 61 (8): 676–680.

Samson, Paul R. and David Pitt, eds. 1999. *The Biosphere and Noosphere Reader: Global Environment, Society and Change*. London and New York: Routledge.

Saraswati, B. 1993. "The implicit philosophy and worldview of Indian tribes." In *Contemporary Philosophy: A New Survey*, Volume 7, *Asian Philosophy*, edited by G. Fløistad, 121–136. Dordrecht: Kluwer Academic Publishers.

Schroeder, Paul. 2018. "Pain sensitivity in fish." *CAB Reviews* 13 (049). http://www.cabi.org/cabreviews

Schweitzer, Albert. 1969. *Reverence for Life* (Translated from the original in German *Strassburger Predigten* © 1966 by Reginald H. Fuller). New York, Evanston, and London: Harper & Row, Publishers. Digitized by the Internet Archive in 2010.

Segobaetso, Benjamin. 2018. "Ethical implications of the colonization, privatization and commercialization of outer space." Master's Research Paper, Ottawa: Saint Paul University. https://ruor.uottawa.ca/bitstream/10393/38318/1/Benjamin_Segobaetso_2018.pdf [Accessed 21 July 2023].

Segundo-Ortin, Miguel, and Paco Calvo. 2022. "Consciousness and cognition in plants." *WIREs Cognition Science* 13: e1578: 1–23. https://doi.org/10.1002/wcs.1578

Sen Gupta, Kalyan. 2005. *The Philosophy of Rabindranath Tagore*. Routledge.

Sessions, George. 1977. "Spinoza and Jeffers on man in nature." *Inquiry* 20 (1–4): 481–528.

Sessions, George. 1987. "The deep ecology movement: A review." *Environmental Review* 11 (2): 105–125.

Sessions, George, ed. 1995. *Deep Ecology for the 21st Century: Readings on the Philosophy and Practice of the New Environmentalism*. Boston: Shambhala.

Singer, Peter. 1974. "All animals are equal." *Philosophic Exchange* 5 (1), Article 6: 103–116. http://digitalcommons.brockport.edu/phil_ex/vol5/iss1/6 [Accessed 07 May 2022].

Singer, Peter. 1975. *Animal Liberation: The Definitive Classic of the Animal Movement*. New York: Fortieth Anniversary Edition, 2015, Iconic E Books from Open Road Media.

Singer, Peter. 1976. "All animals are equal." In *Animal Rights and Human Obligations*, edited by Tom Regan and Peter Singer, 148–162. Englewood Cliffs.: Prentice-Hall.

Singh, L.J., B. Singh, and Abhik Gupta. 2003. "Environmental ethics in the culture of Meeteis from North East India." In *Asian Bioethics in the 21st Century*, edited by S.Y. Song, Y.M. Koo, and D.R.J. Macer, 320–326. Tsukuba: Eubios Ethics Institute.

Skafish, Peter. 2016. "The metaphysics of extra-moderns: On the decolonization of thought – A conversation with Eduardo Viveiros de Castro." *Common Knowledge* 22 (3): 393–414. https://doi.org/10.1215/0961754X-3622248

Smith, Adam. 1790. *The Theory of Moral Sentiments*, Sixth Edition. São Paulo, Brasil: MεtαLibri, 2006.

Smith, Michael B. 1998. "The value of a tree: Public debates of John Muir and Gifford Pinchot." *The Historian* 60 (4): 757–778. https://doi.org/10.1111/j.1540-6563.1998.tb01414.x

Sneddon, Lynne U., and Matthew C. Leach. 2016. "Anthropomorphic denial of fish pain." *Animal Sentience* 3 (28): 2016.035. https://doi.org/10.51291/2377-7478.1048

Sneddon, Lynne U., Robert W. Elwood, Shelley A. Adamo, and Matthew C. Leach. 2014. "Defining and assessing animal pain." *Animal Behavior* 97: 201–212.

Sneddon, Lynne U., Victoria A. Braithwaite, and Michael J. Gentle. 2003. "Novel object test: Examining nociception and fear in the rainbow trout." *The Journal of Pain* 4 (8): 431–440.

Soulé, Michael E. 1985. "What is conservation biology?" *Bioscience* 35 (11), 727–734. The Biological Diversity Crisis.

Sternberg, R.J. 2022. "Human intelligence." *Encyclopedia Britannica*, April 11. https://www.britannica.com/science/human-intelligence-psychology [Accessed 20 August 2022].

St. John, Donald P. 1992. "Whitman's ecological spirituality." *The Trumpeter* 9 (3). https://trumpeter.athabascau.ca/index.php/trumpet/article/view/423 [Accessed 31 July 2023].

Stone, Christopher D. 1972. "Should trees have standing? – Towards legal rights for natural objects." *Southern California Law Review* 45: 450–501.

Strathern, Marilyn. 1987. "Out of context: The persuasive fictions of anthropology." *Current Anthropology* 28 (3): 251–281.

Struik, Paul C., Xinyou Yin, and Holger Meinke. 2008. "Plant neurobiology and green plant intelligence: Science, metaphors and nonsense." *Journal of the Science of Food and Agriculture* 88: 363–370.

Su, K., J. Ren, Y. Qin, Y. Hou, and Y. Wen. 2020. "Efforts of indigenous knowledge in forest and wildlife conservation: A case study on Bulang people in Mangba village in Yunnan province, China." *Forests* 11: 1178. https://doi.org/10.3390/f11111178

Suess, Eduard. 1999. "The face of the earth." In *The Biosphere and Noosphere Reader*, edited by Paul R. Samson and David Pitt, 23. Taylor and Francis e-Library Edition, 2002. London and New York: Routledge.

Swāmi Gambhirānanda, trans. 1957a. *Eight Upaniṣads*, Volume 1 *(Iś ā, Kena, Katha and Taittiriya) with the commentary of Śaṅkarācārya*. Calcutta: Advaita Ashrama (Publication Department). Reprint, Second Revised Edition, November 1991.

Swāmi Gambhirānanda, trans. 1957b. *Eight Upaniṣads*, Volume 2 *(Aitareya, Mundaka, Māndukya & Kārikā, and Praśna) with the commentary of*

Śankarācārya. Calcutta: Advaita Ashrama (Publication Department). Reprint, Ninth Impression, November 1992.

Swami Nikhilananda, trans. 1959. *The Upanishads – Taittiriya and Chhāandogya,* Volume IV. New York: Harper & Brothers Publishers.

Swamy, P.S., M. Kumar, and S.M. Sundarapandian. 2003. "Spirituality and ecology of sacred groves in Tamil Nadu, India." *Unasylva* 54 (213): 53–58.

Syngai, D. 1999. "Sacred groves of Meghalaya" In *Biodiversity: North East India Perspectives,* edited by B. Kharbuli, D. Syiem, and H. Kayang, 70–76. Shillong: North Eastern Biodiversity Research Cell, North Eastern Hill University.

Tagore, Rabindranath. 1931. *The Religion of Man: Being the Hibbert Lectures for 1922.* London: George Allen and Unwin Ltd.

Tagore, Rabindranath. 1950. *Golpoguchho.* Kolkata: Visva Bharati.

Tagore, Rabindranath. 1982. *Rabindra Rachanabali,* Volume II. Pashchimbanga Sarkar (Government of West Bengal).

Tagore, Rabindranath. 1983. *Rabindra Rachanabali,* Volume III. Pashchimbanga Sarkar (Government of West Bengal).

Tagore, Rabindranath. 1985. *Rabindra Rachanabali,* Volume VI. Pashchimbanga Sarkar (Government of West Bengal).

Tagore, Rabindranath. 1990. *Rabindra Rachanabali,* Volume XII. Kolkata: Visva Bharati.

Tagore, Rabindranath. 1994. "The gardener." In *The English Writings of Rabindranath Tagore,* Volume 1, Poems, edited by Sisir Kumar Das, 121. New Delhi: Sahitya Akademi.

Tagore, Rabindranath. 2021. *Prabandha Samogro – Rabindranath Thakur.* Go Bangla Books. https://www.gobanglabooks.com/2021/08/prabandha-samagra-by-rabindranath.html [Accessed 03 August 2023].

Taiz, Lincoln, Daniel Alkon, Andreas Draguhn, Angus Murphy, Michael Blatt, Chris Hawes, Gerhard Thiel, and David G. Robinson. 2019. "Plants neither possess nor require consciousness." *Trends in Plant Science* 24 (8): 677–687.

Taiz, Lincoln, Daniel Alkon, Andreas Draguhn, Angus Murphy, Michael Blatt, Gerhard Thiel, and David G. Robinson. 2020. "Reply to Trewavas *et al.* and Calvo and Trewavas." *Trends in Plant Science* 25 (3): 218–220.

Takakura, H. 2010. "Arctic pastoralism in a subsistence continuum: A strategy for differentiating familiarity with animals." In *Good to Eat, Good to Live with: Nomads and Animals in Northern Eurasia and Africa,* edited by F. Stammler and H. Takakura, 21–42. Sendai: Tohoku University, Centre for Northeast Asia Studies.

Talukdar, Simi, and Abhik Gupta. 2018. "Attitudes towards forest and wildlife, and conservation-oriented traditions, around Chakrashila Wildlife Sanctuary, Assam, India." *Oryx* 52 (3): 508–518.

Tank, Nandini. 2019. "*Binti*: Re-thinking Santal identity through the creation myth." *Lokaratna* XII: 17–28. Bhubaneswar: Folklore Foundation.

Taylor, Bron. 2012. "Wilderness, spirituality and biodiversity in North America – Tracing an environmental history from occidental roots to earth day." In *Wilderness in Mythology and Religion,* edited by Laura Feldt, 293–324. Berlin: De Gruyter.

Taylor, Bron, Guillaume Chapron, Helen Kopnina, Ewa Orlikowska, Joe Gray, and John J. Piccolo. 2020. "The need for ecocentrism in biodiversity conservation." *Conservation Biology* 34 (5): 1089–1096.

Taylor, Paul W. 1981. "The ethics of respect for nature." *Environmental Ethics* 3 (3): 197–218.

Taylor, Paul W. 1986. *Respect for Nature: A Theory of Environmental Ethics,* 25th Anniversary Edition. Princeton and Oxford: Princeton University Press.

Teilhard de Chardin, Pierre. 1966. *The Appearance of Man*. New York: Harper & Row Publishers.

Teilhard de Chardin, Pierre. 1971. *Activation of Energy*. New York and London: A Helen and Kurt Wolff Book, Harcourt Brace Jovanovich.

Teilhard de Chardin, Pierre. 1999a. "The phenomenon of man." In *The Biosphere and Noosphere Reader*, edited by Paul R. Samson and David Pitt, 71–73. Taylor and Francis e-Library Edition, 2002. London and New York: Routledge.

Teilhard de Chardin, Pierre. 1999b. "The antiquity and world expansion of human culture." In *The Biosphere and Noosphere Reader*, edited by Paul R. Samson and David Pitt, 73–79. Taylor and Francis e-Library Edition, 2002. London and New York: Routledge.

The Ecologist. 2016. "Greens commit to rights of nature law." *The Ecologist*, February 29. https://theecologist.org/2016/feb/29/greens-commit-rights-nature -law [Accessed 06 July 2023].

Thomson, James D., Miruna A. Draguleasa, and Marcus Guorui Tan. 2015. "Flowers with caffeinated nectar receive more pollination." *Arthropod-Plant Interactions* 9: 1–7. https://doi.org/10.1007/s11829-014-9350-z

Thoreau, Henry David. 1895. *Poems of Nature*, selected and edited by Henry S. Salt and Frank B. Sanborn. The Project Gutenberg EBook of Poems of Nature, by Henry David Thoreau, 2019 [EBook #59988].

Thoreau, Henry David. 1862. "Walking." *The Atlantic Monthly* IX (LVI): 657–674. The Project Gutenberg eBook of Walking, 1997, [eBook #1022]. https://www .gutenberg.org/files/1022/1022-h/1022-h.htm [Accessed 06 September 2023].

Thoreau, Henry David. 1906. *The Maine Woods*. The Writings of Henry David Thoreau, Volume III (of 20). Boston and New York: Houghton Mifflin and Company. The Project Gutenberg eBook of the Maine Woods, by Henry David Thoreau, 2013 [eBook #42500].

Thoreau, Henry David. 1995. *Walden or Life in the Woods, and on the Duty of Civil Disobedience*. The Project Gutenberg EBook of Walden, by Henry David Thoreau [eBook #205].

Toko, K., M. Souda, T. Matsuno, and K. Yamafuji. 1990. "Oscillations of electrical potential along a root of a higher plant." *Biophysics Journal* 57: 269–279.

Trewavas, Anthony. 2017. "The foundations of plant intelligence." *Interface Focus* 7: 20160098. http://doi.org/10.1098/rsfs.2016.0098

Tucker, M.E. 2001. "Confucianism and deep ecology." In *Deep Ecology and World Religions*, edited by D.L. Barnhill and R.S. Gottlieb, 127–152. Albany: State University of New York Press.

Tylor, Edward Burnett. 1871. *Religion in Primitive Culture*. First Harper Torchbook Edition Published 1958. New York: Harper and Brothers Publishers.

United Nations. 1973. *Report of the United Nations Conference on the Human Environment*. Stockholm, June 5–16, 1972. New York: United Nations.

United Nations. 1982. *World Charter for Nature*. UN General Assembly (37th Sess: 1982–1983). https://digitallibrary.un.org/record/39295?ln=en [Accessed 05 July 2023].

United Nations. 1987. *Our Common Future: Report of the World Commission on Environment and Development*. Transmitted to the General Assembly as an Annex to document A/42/427 - Development and International Cooperation: Environment.

United Nations. 1992a. "Report of the United Nations Conference on Environment and Development (Rio de Janeiro 3–14 June 1992)." United Nations General Assembly A/CONF.151/26 (Vol. I). https://www.un.org/en/development/desa/ population/migration/generalassembly/docs/globalcompact/A_CONF.151_26 _Vol.I_Declaration.pdf [Accessed 15 August 2022].

United Nations. 1992b. *Convention on Biological Diversity.* https://www.cbd.int/doc/legal/cbd-en.pdf [Accessed 06 July 2023].

United Nations. 2012. *The Future We Want.* Outcome Document of the United Nations Conference on Sustainable Development, Rio de Janeiro, Brazil, 20–22 June, 2012. Rio + 20 United Nations Conference on Sustainable Development. https://sustainabledevelopment.un.org/content/documents/733FutureWeWant .pdf [Accessed 06 July 2023].

United Nations. 2023. *Harmony with Nature: Chronology.* www.harmonywith natureun.org/chronology/ [Accessed 06 July 2023].

United Nations Office for Outer Space Affairs (UNOOSA). 2008. *United Nations Treaties and Principles on Outer Space and Related General Assembly Resolutions.* United Nations Publication, Sales No. E.08.I.10. New York: United Nations. https://www.unoosa.org/pdf/publications/st_space_11rev2E.pdf [Accessed 17 July 2023].

United States Geological Survey (USGS). 2023. *Mendenhall Research Fellowship Program: Evaluating Mineral Resources on Mars for Exploration and Colonization.* https://www.usgs.gov/centers/mendenhall-research-fellowship -program/18-27-evaluating-mineral-resources-mars-exploration [Accessed 17 July 2023].

URL 1. https://www.dictionary.com/browse/ecocentrism [Accessed 12 September 2023].

URL 2. https://study.com/learn/lesson/ecocentric-biocentric-philosophies-definition -examples.html [Accessed 12 September 2023].

URL 3. *Jainism Simplified Chapter 3 – Jiva (Living Beings).* http://umich.edu/ ~umjains/jainismsimplified/chapter03.html [Accessed 30 September 2021].

URL 4. *Jiva and Ajiva* (Soul and Lifeless Substances). www.jainbelief.com/PPOJ/15 .htm [Accessed 30 September 2021].

US Department of State, and NASA. 2020. *The Artemis Accords: Principles for Cooperation in the Civil Exploration and Use of the Moon, Mars, Comets, and Asteroids for Peaceful Purposes.* https://www.nasa.gov/specials/artemis-accords /img/Artemis-Accords-signed-13Oct2020.pdf [Accessed 19 July 2023].

Van den Born, R.J.G. 2007. *Thinking Nature: Everyday Philosophy of Nature in The Netherlands.* Ph.D. Thesis, Radboud University, Nijmegen, The Netherlands.

Vernadsky, Vladimir I. 1998. *The Biosphere.* New York: Springer.

Vernadsky, Vladimir I. 1999. "Geochemistry." In *The Biosphere and Noosphere Reader,* edited by Paul R. Samson and David Pitt, 26–27. Taylor and Francis e-Library Edition, 2002. London and New York: Routledge.

Vernadsky, Vladimir I. 2000–2001. "Problems of biogeochemistry II." *21st Century* (Winter 2000–2001): 20–39.

Vernadsky, W.I. 1945. "The biosphere and the noösphere." *American Scientist* 33 (1): 1–12.

Vitebsky, P., A. Alekseyev. 2015. "What is a reindeer? Indigenous perspectives from northeast Siberia." *Polar Record* 51 (259): 413–421. https://doi.org/10.1017/ S0032247414000333

Viveiros De Castro, E. 1998. "Cosmological deixis and Amerindian perspectivism." *Journal of Royal Anthropological Institute* 4: 469–488.

Viveiros de Castro, E., Walford, A. 2011. "Zeno and the art of anthropology: Of lies, beliefs, paradoxes, and other truths." *Common Knowledge* 17 (1): 128–145.

Voltaire. 1976. "A reply to Descartes." In *Animal Rights and Human Obligations,* edited by Tom Regan and Peter Singer, 67–68. Englewood Cliifs: Prentice-Hall, Inc. Originally published in H.I. Woolf, selected and trans., *Voltaire, The Philosophical Dictionary*: "Animals" (New York: Knopf, 1924).

Wall, Mike. 2019a. "Bill Nye: It's space settlement, not colonization." *Space*, October 25. https://www.space.com/bill-nye-space-settlement-not-colonization .html [Accessed 16 October 2022].

Wall, Mike. 2019b. "Looks like Elon Musk is serious about nuking Mars." *Space*, August 22. https://www.space.com/elon-musk-serious-nuke-mars-terraforming .html [Accessed 23 July 2023].

Walters, E.T. 2016. "Pain-capable neural substrates may be widely available in the animal kingdom." *Animal Sentience* 3 (37). https://doi.org/10.51291/2377-7478 .1067

Washington, H., B. Taylor, H. Kopnina, P. Cryer, and J.J. Piccolo. 2017. "Why ecocentrism is the key pathway to sustainability." *The Ecological Citizen* 1: 35–41.

Wei-Ming, T. 1998. "Beyond the enlightenment mentality." In *Confucianism and Ecology*, edited by M.E. Tucker and J. Berthrong, 3–21. Cambridge: Harvard University Press.

Whanganui District Council. 2023. *Te Awa Tupua – Whanganui River Settlement*. Whanganui District Council. https://www.whanganui.govt.nz/ About-Whanganui/Our-District/Te-Awa-Tupua-Whanganui-River-Settlement [Accessed 13 July 2023].

White, Lynn Jr. 1967. "The historical roots of our ecological crisis." *Science* 155 (3767): 1203–1207. https://doi.org/10.1126/science.155.3767.1203

White, Peter S. 2013. "Derivation of the extrinsic values of biological diversity from its intrinsic value and of both from the first principles of evolution." *Conservation Biology* 27 (6): 1279–1285.

Whitman, W. 1855–1892. *Leaves of Grass*. The Project Gutenberg EBook of Leaves of Grass, 1998 [EBook # 1322]. https://www.gutenberg.org/cache/epub/1322/ pg1322.txt [Accessed 25 October 2023].

Whitman, W. 2005. *Complete Prose Works*. The Project Gutenberg EBook of Complete Prose Works, by Walt Whitman [EBook #8813]. https://www .gutenberg.org/cache/epub/8813/pg8813.txt [Accessed 25 October 2023].

Willerslev, R. 2011. "Frazer strikes back from the armchair: A new search for the animist soul." Malinowski Memorial Lecture 2010. *Journal of the Royal Anthropological Institute (N.S.)* 17: 504–526.

Willerslev, R. 2013. "Taking animism seriously, but perhaps not too seriously?" *Religion in Society: Advances in Research* 4: 41–57.

Willerslev, R., and O. Ulturgasheva. 2012. "Revisiting the animism versus totemism debate: Fabricating persons among the Eveny and Chukchi of North-eastern Siberia." In *Animism in Rainforest and Tundra: Personhood, Animals and Things in Contemporary Amazonia and Siberia*, edited by M. Brightman, V.E. Grotti, and O. Ulturgasheva, 48–68. New York and Oxford: Berghahn Books.

Wilson, E.O. 1988. *Biophilia*. Cambridge: Harvard University Press.

Wittenburg, Nicole Elaine. 2012. *The Bean Field Quandary: Environmental Ethics in Emerson and Thoreau*. Master of the Arts Thesis. Montclair State University, Montclair.

Wong, P-H. 2015. "Confucian environmental ethics, climate engineering, and the 'playing god' argument." *Zygon* 50 (1): 28–41.

Woodlief, Ann. 1990. "Emerson and Thoreau as American prophets of eco-wisdom." *Paper Presented to Virginia Humanities Conference, 1990.* https:// transcendentalism.tamu.edu/ecotran

World Commission on Environment and Development (WCED). 1987. *Report of the World Commission on Environment and Development: Our Common Future.* https://sustainabledevelopment.un.org/content/documents/5987our-common -future.pdf

Worster, Donald. 1980. "The intrinsic value of nature." *Environmental Review* 4 (1): 43–49.

Wroth, David. 2020. "Rainbow serpent dreamtime story." https://japingkaaborigi nalart.com/articles/rainbow-serpent/ [Accessed 16 September 2022].

Yano, K. 2008. "Sacred mountains where being of 'Kami' is found." International Council on Monuments and Sites (ICOMOS), Quebec, Canada. http://openarchive.icomos.org/id/eprint/52/1/77-Epvp-23.pdf [Accessed 30 September 2021].

Yarlagadda, Shreya. 2022. "Economics of the stars: The future of asteroid mining and the global economy." *Harvard International Review*, April 08. https://hir .harvard.edu/economics-of-the-stars/ [Accessed 18 July 2023].

Ziltener, Claude. 2007. "The death-hymn of the perfect tree: Metaphor, metamorphosis and the sublimity of music in R.W. Emerson's poems 'Woodnotes I & II'." *SPELL (Swiss Papers in English Language and Literature)* 20: 47–68.

Zimmerman, Michael E. 1993. "Heidegger, Buddhism, and deep ecology." In *The Cambridge Companion to Heidegger*, edited by Charles B. Guignon, 240-XXX. Cambridge: Cambridge University Press.

Zimmerman, Michael E. 2003. "Heidegger's phenomenology and contemporary environmentalism." In *Eco-Phenomenology: Back to the Earth Itself*, edited by Charles S. Brown and Ted Toadvine, 73–102. Albany: State University of New York Press.

Zimmerman, M.J., and B. Bradley. 2019. "Intrinsic vs. extrinsic value." In *The Stanford Encyclopedia of Philosophy*, edited by Edward N. Zalta (Spring 2019 Edition). https://plato.stanford.edu/archives/spr20

Zweers, W. 1994. "Radicalism or historical consciousness: On breaks and continuity in the discussion of basic attitudes." In *Ecology, Technology and Culture; Essays in Environmental Philosophy*, edited by W. Zweers and J.J. Boersema, 63–71. Cambridge: The White Horse Press.

INDEX

For Product Safety Concerns and Information please contact our EU
representative GPSR@taylorandfrancis.com
Taylor & Francis Verlag GmbH, Kaufingerstraße 24, 80331 München, Germany

www.ingramcontent.com/pod-product-compliance
Lightning Source LLC
Chambersburg PA
CBHW060252220326
41598CB00027B/4073

* 9 7 8 1 0 3 2 7 7 0 7 6 5 *